SpringerBriefs in Microbiology

For further volumes:
http://www.springer.com/series/8911

Peter Zuber

Function and Control of the Spx-Family of Proteins Within the Bacterial Stress Response

 Springer

Peter Zuber
Division of Environmental and
    Biomolecular Systems
Institute of Environmental Health
Oregon Health & Science University
Beaverton, OR
USA

ISSN 2191-5385   ISSN 2191-5393   (electronic)
ISBN 978-1-4614-6924-7   ISBN 978-1-4614-6925-4   (eBook)
DOI 10.1007/978-1-4614-6925-4
Springer New York Heidelberg Dordrecht London

Library of Congress Control Number: 2013936518

Printed on acid-free paper

Springer is part of Springer Science+Business Media (www.springer.com)

Contents

Function and Control of the Spx-Family of Proteins Within the Bacterial Stress Response

Introduction

A complex regulatory network governs the microbial response to encounters with harsh environmental conditions. Such networks also influence essential decision-making processes that coordinate metabolic pathways and maintain cellular ultra-structure. The study of such networks has been a fruitful area of research over the years, and continues to be a major focus of biomedical, applied, and environmental studies. Investigations using model bacterial systems revealed conserved mechanisms that neutralize harmful conditions by destroying toxic agents or alleviating their damaging effects. Molecular genetic studies of the model enteric bacterium, *Escherichia coli*, uncovered complex signal transduction, and response systems that govern the processes of toxicity/damage control. Many of these are based on conserved regulatory architecture involving two-component signal transduction systems, alternative RNA polymerase sigma subunits, conserved members of transcription factor families, and post-translational control through modification and proteolysis. Explorations into the Gram-positive bacterial stress response systems have reinforced the importance of conserved regulatory factors, but have also uncovered quite novel players functioning as nodes in the stress management network. Such nodes are the targets of multiple stress-induced regulatory devices and upon stimulation exert their effects on a variety of response effectors. Among the factors that have emerged as important contributors to stress response control in the low G+C Gram-positive bacteria are members of the Spx family of proteins. Spx of *Bacillus subtilis* has been shown to exert direct transcriptional control by interaction with RNA polymerase, by which it directs expression of a battery of genes whose products function in the oxidative stress response and likely other stress response processes. Control of Spx activity is exerted in part through a thiol/disulfide switch that enables Spx to undergo activation directly upon reaction with toxic oxidants. Its concentration is also controlled at the levels of gene transcription and post-translationally through proteolysis catalyzed by ATP-dependent proteases. Transcriptional control of *spx* in the Gram-positive spore-forming

P. Zuber, *Function and Control of the Spx-Family of Proteins Within the Bacterial Stress Response*, SpringerBriefs in Microbiology, DOI: 10.1007/978-1-4614-6925-4_1,
© The Author(s) 2013

bacterium, *Bacillus subtilis* is especially complex, involving three forms of RNA polymerase and at least two transcriptional repressors. Signal transduction systems connected to cell envelope stress transcriptionally control *spx* genes in orthologous Gram-positive systems. High levels of Spx not only lead to transcriptional activation, but also repression of genes functioning in complex developmental programs, such as sporulation and the development of genetic competence (see below). Accumulating evidence indicates that Spx represents a point of convergence to which signals derived from a variety of stress conditions are directed to enact transcriptional and post-transcriptional control. Recent studies have linked these stress conditions, including encounters with clinically relevant antibiotics, to the generation of reactive oxygen species (ROS) and the induction of genes belonging to oxidative stress stimulons. The positioning of Spx at a stress-signaling intersection may represent a preemptive strategy to mobilize defenses against ROS that are generated as a secondary consequence of encounters with toxic agents and harsh environmental conditions.

Spx is a direct participant in the control of gene expression in response to stress and is necessary to render cells resistance to toxic oxidants as well as the lethal conditions affecting the cell envelope of certain species. The importance of Spx is further highlighted by the demonstration that deletion of *spx* paralogous genes is lethal in some members of *Firmicutes*. Heightened Spx activity in *B. subt*ilis results in the activation of a large regulon of over 140 operons and in elevated resistance to toxic agents. Because Spx can exert a profound effect on the operation of complex, global regulatory systems, the overview of Spx-dependent regulation will be followed with a summary of stress-induced global gene regulation in Gram-positive microorganisms in which functional Spx is produced. The remainder of the review will highlight the research aimed at uncovering the regulation and function of Spx with emphasis on its relationship with other stress-induced control systems.

An Overview of Spx

Orthologs of Spx can be found in low GC Gram-positive bacteria, but the family of proteins to which it belongs encompasses a much wider spectrum of bacteria (Turlan et al. 2009). Spx is a member of the ArsC family, whose founding member is the arsenate reductase encoded by the plasmid R775 that replicates in *E. coli* (Martin et al. 2001; Zuber 2004). Crystallographic studies have confirmed the conservation of 3D structure of the ArsC and Spx proteins (Newberry et al. 2005). Two domains can be discerned from inspection of the structure, with the N-and C-terminal segments constituting one domain, and the central segment of the protein forming the other (Fig. 1). The N- and C-terminal regions of Spx/ArsC are linked by a 4-stranded beta sheet. In ArsC, the domain composed of the N-and C-terminal regions contains the active site Cys residue that forms a covalent link with arsenate, thus generating the reaction intermediate. This Cys

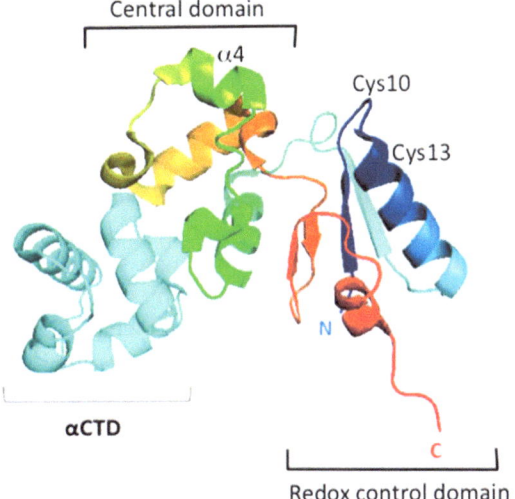

Fig. 1 Structure of Spx bound to RNA polymerase alpha C-terminal domain (*pale blue*). Two linker polypeptide coils link the redox control domain with the central domain. Red indicates the C-terminal portion of Spx that functions in Spx proteolytic control. Also shown is the alpha 4 helix that functions in promoter DNA recognition, and the two Cys residues that constitute the thiol/disulfide redox center. The N- and C-terminal regions constitute the redox control domain are held together by a 4-stranded beta sheet. (Adopted from Newberry et al. 2005)

residue is conserved among most of the Spx orthologs, where it is often partnered with another Cys residue to form a redox disulfide reactive center. Hence, the domain made up of the N- and C-terminal regions is referred to here as the redox control domain (Fig. 1).

A bioinformatic analysis was carried out by Turlan et al. (2009) to uncover the differences among the ArsC family members, which resulted in the establishment of four clades (bootstrap values between 92 and 98) of the ArsC/Spx homologs. *Firmicutes* proteins constituted two of the clades, which included a group of ArsC-like proteins, including Spr1316 of *Streptococcus pneumoniae* and YusI of *B, subtilis*, the gene for which resides in an Spx-controlled operon required for lipoate synthesis (Christensen et al. 2011). The other clade containing *Firmicutes* ArsC-family members is heavily represented by the transcriptional regulator SpxA from *Listeria monocytogenes*, *Streptococcus mutans*, *S. pyogenes*, *Staphylococcus aureus*, *Lactococcus lactis*, *the B. subtilis,* and the *B. cereus/thuringiensis/anthracis* groups. The third clade is represented by the YffB protein and its orthologs from *Brucellaceae* (Buchko et al. 2011) and other members of the alphaproteobacteria and a few members of gammaproteobacteria (*Pseudomonas*) (Teplyakov et al. 2004). Some are annotated as the "Spx/MgsR (modulator of general stress response)" group, but these are of unknown function. Confirmation of the ArsC-like structure of this third group was established from the crystal structure of the YffB proteins of *B. melitensis* and *P. aeruginosa*

(Buchko et al. 2011; Teplyakov et al. 2004). The fourth and largest clade is the "ArsC–ArsC" grouping that includes the bona fide arsenate reductase from *E. coli* and its orthologs from a variety of Gram-positive and -negative species.

Mutations in a gene encoding an ArsC-like protein were reported in *Lactococcus lactis* (Duwat et al. 1999). The mutation elevated temperature resistance in a *recA* mutant, stimulated proteolysis, and enhanced stress tolerance. The gene is called *trmA*, but this assignment had already been given to the gene encoding a highly conserved tRNA-rRNA methylase in bacteria, including members of the *Firmicutes*. In *B. subtilis*, the gene, *yjbD* (now known as *spx*), encoding an ArsC-like protein, was observed to be activated by several stress-inducing conditions, including phosphate starvation and elevated temperature (Antelmann et al. 2000; Petersohn et al. 2001). The *spx* gene resides within a dicistronic operon with *yjbC* that is controlled by several minor forms of RNA polymerase σ subunit, including σ^M, which was observed to stimulate *yjbCspx* transcription (Jervis et al. 2007; Thackray and Moir 2003). An investigation into the role of the ATPase, ClpX in sporulation and competence gene expression uncovered mutations in *spx* (suppressor of *clpP* and *clpX*) of *B. subtilis* (Nakano et al. 2001, 2002). The ClpX protein is a subunit of the ATP-dependent protease ClpXP, where ClpX is the substrate-binding hexameric unfoldase that delivers the substrate protein to the proteolytic chamber formed by two heptameric rings of ClpP subunits (Sauer and Baker 2011). Mutations in *clpX* or *clpP* cause severe defects in sporulation, competence development, and growth in synthetic minimal media (Liu et al. 1999; Liu and Zuber 2000; Nakano et al. 2003b). These defects were largely alleviated by second site mutations in *spx*. Mutations with a similar *clpX*-suppressing phenotype were discovered in the *rpoA* gene encoding the RNA polymerase α subunit of *B. subtilis* (Nakano et al. 2000), and one of these mutations was later shown to be in a codon specifying an amino acid within the contact interface between Spx and the C-terminal domain of the α subunit (Newberry et al. 2005).

The *clpX* phenotype was attributable to the accumulation of Spx, which is a substrate for ClpXP, and its interaction with RNA polymerase (Nakano et al. 2003b). High concentrations of Spx lead to interference with the interaction between RNA polymerase and transcriptional regulators due to the occupation of regulator binding surfaces on the α subunit by Spx (Zhang et al. 2006). Recent data suggested that repression by Spx could also be exerted by a different mechanism, possibly involving Spx-DNA interaction (Rochat et al. 2012). Transcriptomic studies showed that Spx at high concentrations activates the transcription of genes that function in cysteine production and in thiol homeostasis (Nakano et al. 2003a). Null mutations of *spx* and a mutation of *rpoA* that disrupted Spx-RNA polymerase interaction conferred hypersensitivity to the thiol-specific oxidant, diamide. However, the Spx regulon is large and several of the genes that are activated greater than 3-fold are involved in other functions, including cell wall metabolism (Zuber et al. 2011). Recent chromatin immunoprecipitation studies using epitope-tagged Spx have uncovered genes directly contacted by Spx/RNA polymerase complex. While these data largely confirmed previous transcriptomic work, more Spx regulon members were uncovered, including the gene encoding

ClpX (Rochat et al. 2012). In all, 257 genes were identified as being potentially affected by Spx-RNA polymerase interaction.

In *B. subtilis*, Spx concentration is regulated at several levels. Transcription is regulated through several forms of RNA polymerase bearing alternative σ subunits (Cao et al. 2002b). Spx is under proteolytic control involving ClpXP protease and a substrate recognition factor, as discussed below. Finally, Spx activity is under redox control, as productive interaction with RNA polymerase requires formation of an intramolecular disulfide at the redox CXXC center (Nakano et al. 2005). Although Spx has been described as a regulator of the oxidative stress response, the fact that its production is induced by multiple stress conditions and its regulon encompasses genes specifying functions not directly related to the oxidative stress response suggest that Spx might participate more globally in the Gram-positive stress response.

To begin to address the position Spx occupies in the stress response network, the next section will provide a summary of the regulatory circuits governing stress response mechanisms in Gram-positive organisms that produce orthologous Spx proteins. The following is not an in-depth treatment of the subjects, and the reader is directed to several recent reviews (Gaup et al. 2012; Higgins and Dworkin 2012; Jordan et al. 2008; Price 2011; Zuber 2009) that cover the subjects in more detail.

Regulation of the Stress Response in Gram-Positive Bacteria

Members or the phylum, *Firmicutes*, represent a large family of Gram-positive bacterial species that share many common features within their regulatory networks that orchestrate stress responses. Most of the low G+C Gram-positive bacteria relevant to this review are members of *Firmicutes*. The species affiliated with *Firmicutes* include those within the orders *Bacillales* (Genera: *Bacillus*, *Listeria*, *Staphylococcus*), *Lactobacillales* (*Lactobacillus*, *Enterococcus*, *Streptococcus*), *Clostridiales* (*Clostridium*, *Eubacterium*, *Peptococcus*), and *Erysipelotrichi* (*Catenibacterium*, *Coprobacillus*). Members of the *Firmicutes* include notable pathogens and industrially important bacteria used in the large-scale production of enzymes and antibiotics. The Gram-positive bacteria are very diverse, encompassing species that inhabit many environments ranging from soil, fresh and saltwater, as well as on and within multicellular host organisms. They exhibit aerobic, anaerobic, or facultative lifestyles and conduct fermentative and respiratory metabolic processes. While members of the phylum inhabit wide-ranging environments, some are restricted to narrow ecological spaces. Certain members of *Erysipelotrichi*, *Clostridia*, and *Lactobacillales* are found within the oral and/or gut microbiomes (Peris-Bondia et al. 2011). All possess a three-dimensional mesh of peptidoglycan that envelops a single cytoplasmic membrane. While similarities in ultrastructure characterize the phylum, there is much morphological diversity, even within orders. The *Bacillales* include the spore-forming, rod shape, motile

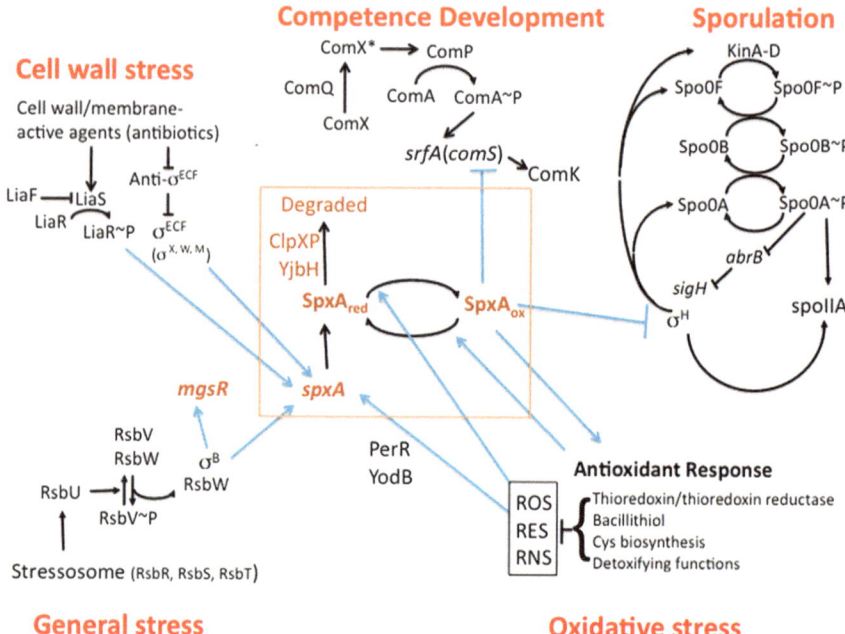

Fig. 2 Relationship of Spx with control circuits governing developmental programs and stress responses in Firmicutes. *Blue arrows* (*pointed* and *flat*) represent connections between spx expression and activity with the developmental and stress response systems indicated in *red letters*

bacteria within the genus *Bacillus* and the nonspore-forming, nonmotile coccus, *Staphylococcus*. The *Clostridiales* also include spore-forming anaerobes as well as nonspore-forming, anaerobic cocci (*Peptostreptococcus*). Despite these differences, genomic analyses have shown remarkable similarities in primary gene structure and close synteny in gene organization.

Figure 2 shows a schematic diagram of the position Spx occupies in the stress response network. The figure does not include all of the potential interactions Spx can establish with components of the regulatory network. The figure conveys the different ways Spx production and activity is influenced by response regulatory systems and how Spx exerts control over systems of differentiation and stress resistance.

Control Within Complex Developmental Programs and the Negative Impact of Spx

The diverse lifestyles, metabolic capabilities, and the complex developmental cycles in several members of the *Firmicutes* would suggest significant divergence in regulatory network components and architecture. Inspection of genome composition and review of studies into gene regulation, to some degree, support this conclusion. Several members of the *Bacillales*, as noted earlier, can undergo sporulation in response to nutritional stress. This is a process of cellular differentiation in which a vegetative cell transitions into a prespore state in which the cell is composed of two compartments, one of which will become the dormant spore and the larger compartment, or mother cell that supports spore development, but ultimately undergoes lysis when the mature spore is released. Genetic control of the process operates at several levels and requires fundamental changes to the gene expression machinery (de Hoon et al. 2010; Piggot and Hilbert 2004). Most notable among these are the spore-specific RNA polymerase forms bearing unique σ subunits (Stragier and Losick 1990). The assembly of alternative RNA polymerase forms, bearing spore-specific σ subunits, is the result of a signal-transducing phosphorelay responsive to the nutritional environment, and compartment-specific control of gene expression responsive to intercompartmental (mother cell-forespore) communication (Duncan et al. 1994; Hoch 1993; Stragier and Losick 1996). The process is relevant to Spx since transcriptomic analysis of an *spx* null mutant provided evidence for elevated sporulation gene transcription in a minimal medium that does not promote spore formation (Zuber et al. 2011). This would suggest that some oxidized Spx is present during the transition from exponential to stationary phase of the growth curve, where it could hinder transcriptional control of sporulation genes through RNA polymerase interaction. This is consistent with earlier work that demonstrated a requirement for ClpX in sporulation-specific gene transcription that could be overcome by mutations affecting Spx-RNA polymerase interaction (Liu et al. 1999; Nakano et al. 2000).

The master regulator of sporulation is Spo0A, a member of the response regulator family of proteins that undergoes sequential levels of phosphorylation over time to reach a level of activity that triggers sporulation initiation (Fujita and Losick 2005; Hoch 2000). This requires a set of sensory histidine kinases, KinA-D (Banse et al. 2011; Jiang et al. 2000) that deliver phosphate to an intermediary response regulator, Spo0F. The phosphate is rapidly transferred via the phosphotransferase, Spo0B to Spo0A (Burbulys et al. 1991). Spo0A, thus activated, stimulates transcription of early sporulation genes by direct interaction with RNA polymerase (Baldus et al. 1995; Seredick and Spiegelman 2007). Key to this interaction is the alternative RNA polymerase sigma subunit, σ^H, which is required for optimal transcription of the *kinA*, *spo0F,* and *spo0A* genes (Predich et al. 1992), increases in concentration by a Spo0A-dependent regulatory mechanism and accumulates during stationary phase of the growth curve (Fujita and Losick 2005).

Sequentially higher levels of phosphorylated Spo0A (Spo0A~P) activate specific sets of genes, either through direct activation of transcription initiation (Fujita and Losick 2005) or by repressing the gene *abrB* (Strauch et al. 1990), whose product globally represses genes that are normally induced during the transition from vegetative growth to growth restriction resulting from nutrient depletion (Strauch 1993). Low concentrations of Spo0A~P are required for cells to make the adjustment into the transition state and to select an appropriate physiological choice such as competence development and establishment of a motile state

(Fujita et al. 2005) depending on the conditions encountered. Higher concentrations of Spo0A~P during prolonged exposure to nutrient depletion, trigger the expression of sporulation-specific genes through direct transcriptional activation. The genes activated by Spo0A, along with the σ^H form of RNA polymerase, are those residing in the *spoIIA* operon that specifies the sporulation sigma subunit, σ^F and proteins functioning in σ^F control (Baldus et al. 1995). Thus, the graded increase in Spo0A~P concentration leads the cell through the transition state of growth restriction to the developmental program of sporulation.

A further layer of control is enacted through post-translational control of RNA polymerase σ subunits and proteolytic processing of inactive pro-sigma proteins to yield active subunits. The mechanisms of sigma subunit activation are at the core of cellular compartment (forespore/mother cell)-specific gene expression. Thus, the early sporulation sigma subunit, σ^F, directs transcription of genes within the forespore compartment, but one of the products of the σ^F regulon activates transcription in the mother cell by stimulating the activity of a membrane-bound protease that processes an inactive pro-σ^E protein to its transcriptionally active form that operates within the mother cell (Hofmeister et al. 1995; Karow et al. 1995). In this way, development in the two compartments proceeds in a coordinated way to ensure proper assembly of the prespore structures. Sigma subunit activation and proteolytic processing control are triggered by morphological cues, such as formation of the sporulation septum that separates the forespore from the mother cell compartment (Errington 1993; Stragier and Losick 1990).

The *spx* null mutant shows elevated transcripts from genes that are activated early in sporulation, specifically those requiring the σ^E, σ^H, and σ^F forms of RNA polymerase (Zuber et al. 2011). Mutants bearing a null mutation of *clpX* show reduced σ^H-dependent transcription that can be alleviated by mutations in *spx* (Liu et al. 1999; Liu and Zuber 2000; Nakano et al. 2001). It is currently unclear how Spx exerts a negative effect on sporulation gene expression; whether repression is due to its RNA polymerase-binding activity, indirect metabolic effects of Spx-directed transcription or both. However, a mutation in *rpoA* affecting Spx interaction with RNA polymerase also alleviates the sporulation negative phenotype of the *clpX* mutant (Nakano et al. 2000), implying that Spx interaction with RNA polymerase negatively affects sporulation-specific transcriptional control.

Quorum-Sensing Control of Competence Development and Virulence in Firmicutes

The control of genetic competence development in *B. subtilis*, which has been studied in great detail (Grossman 1995; Hamoen et al. 2003; Tortosa and Dubnau 1999), is also a target of Spx negative control. Competence is a physiological state in which a cell is able to take up DNA from the environment and incorporate the internalized DNA into its genome by recombination. The global transcription factor, ComK, interacts with the regulatory regions of genes whose products function in DNA uptake and in establishing the "K state", a semidormant physiological condition that characterizes competent cells (Berka et al. 2002; Maamar and Dubnau 2005). Only a fraction of the cell population undergoes competence development in a culture, as transient elevations within the "noise" of ComK levels cause occasional stochastic transitions to the competence pathway (Maamar et al. 2007). The gene encoding the ComK ortholog is present in *Listeria* as is the ortholog of MecA (Borezee et al. 2000; Rabinovich et al. 2012), which functions as a substrate recognition factor for the proteolysis of ComK by the ATP-dependent protease, ClpCP in noncompetent cells. That the regulatory components are shared with *Listeria monocytogenes* that is not known to undergo competence development suggests that they might serve different roles in the competent and noncompetent species. The genomes of several strains isolated from food products bear an interrupted allele of *comK*, due to the integration of the prophage form of phage A118 (Verghese et al. 2011). The importance of *comK* in virulence was demonstrated recently in a study that uncovered a novel prophage excision switch that creates a functional *comK* allele in cells within an infected host. The product of the reconstructed *comK* allele drives expression of genes required for phagosomal escape during Listerial infection (Rabinovich et al. 2012).

To activate ComK in *B. subtilis*, a two-component signal transduction system (TCSTS), ComA-ComP, functioning, respectively, as response regulator and histidine protein kinase, activate the transcription of a gene specifying a small protein that inhibits MecA-dependent proteolysis of ComK (Grossman 1995; Turgay et al. 1997). This TCSTS is at the heart of a quorum sensing mechanism mediated by a secreted modified peptide, ComX (Magnuson et al. 1994). Conditions of nutritional stress also stimulate ComAP activity (Solomon et al. 1995). The sensory domain of ComP protrudes from the cell and can contact the extracellular ComX peptide to trigger kinase activity and subsequent ComA phosphorylation. The DNA encoding ComX, the modifying enzyme and the sensory domain of ComP constitute a genetic module of linked sequence elements that is the site of genetic polymorphism (Tran et al. 2000; Tortosa et al. 2001). The variability in sequence modules between *B. subtilis* isolates evolved from specifying selection and highlights the strain-specific nature of the peptide-sensor relationship that operates the quorum sensing system of control. The genes encoding the peptide, its modifying enzyme, and the ComAP TCSTS reside in an operon that includes the polymorphic module, and the organization of this operon is conserved in other members of *Firmicutes*, notably *comCDE* in *Streptococcus* and *agrBCDA* in *Staphylococcus aureus* (Tortosa and Dubnau 1999). The functional analog in *B. subtilis*, ComA, upon activation, triggers transcription initiation of *srfA* (Nakano and Zuber 1991; Weinrauch et al. 1990), a large operon encoding the enzyme complex that catalyzes surfactin biosynthesis, and the small protein ComS with inhibits ComK degradation by MecA/ClpCP, resulting in ComK accumulation (D'Souza et al. 1994; Hamoen et al. 1995; Turgay et al. 1997).

Such systems are susceptible to Spx negative control, as it has been demonstrated that interaction between Spx and RNA polymerase blocks ComA-activated transcription by interfering with ComA-RNA polymerase interaction (Nakano et al. 2003a, b, Zhang et al. 2006). Spx also represses ComK activated transcription in vitro (Nakano, S., and P. Z., unpublished). Spx also exerts negative control over competence development in *S. pneumoniae* (Turlan et al. 2009), although the exact mechanism is not known at this time. As with sporulation, the notion is that Spx, when it is activated under oxidative stress conditions, delays complex developmental programs until redox homeostasis is restored.

Cell Envelope Stress Response Systems that Promote *spx* Expression

The complex developmental pathways are energy intensive and are thought to be the last choice among many responses bacteria can mobilize when encountering stress (Reder et al. 2012a, b, Gonzalez-Pastor et al. 2003). There are more focused responses to environmental insult that target specific structures or metabolic and repair pathways. For example, responses to chemical/physical attacks upon the cell envelope result in changing the composition of the cell wall and/or its charge. Nevertheless, even in these cases, the response might be accompanied by wide-ranging changes in the bacterial transcriptome (Jordan et al. 2008). The *spx* gene is transcriptionally activated by factors that function in the response to cell envelope stress (Fig. 2), creating the potential for genome-wide transcriptional changes.

The three-dimensional mesh of crosslinked glycan chains that constitutes the peptidoglycan is an essential structure that must be maintained, remodeled, and augmented in order to accommodate growth and survival (Silhavy et al. 2010). Another essential component of the Gram-positive cell envelope is teichoic acid (TA), which is attached to the peptidoglycan (wall TA), or to the cytoplasmic membrane (lipoTA) (Formstone et al. 2008; Minnig et al. 2005; Perego et al. 1995). Formation of the cell wall proceeds with the synthesis of the N-acetylglucosamine-N-acetylmuramic acid moiety linked to the pentapeptide linker, which is the unit of peptidoglycan assembly that is translocated to the exterior. This translocation is mediated by a carrier lipid (lipid II, or undecaprenyl pyrophosphate), and is followed by transglycosylation linking the cell wall assembly unit to the growing peptidoglycan chain. All of the steps in cell wall synthesis are essential reactions, and it is no surprise that many antimicrobial agents, natural or synthetic, target either the enzymes that catalyzes these reactions or their substrates. Nearly every step is the target of one or more antibiotics with bacteriocidal activity (Jordan et al. 2008), several of which are currently in clinical use to treat infections. There is evidence that the antibiotic-induced inhibition of cell envelope synthesis and maintenance generates secondary damaging effects due to ROS generation (Kohanski et al. 2007, 2010; Mols and Abee 2011). Studies presented in the recent literature have implicated oxidative stress as the bacteriocidal event that is induced by antimicrobial agents that target the cell wall (Kohanski et al. 2007, 2010). Recent reports, however refute these assertions, showing that antibiotics, in clinically relevant concentrations, are equally bactericidal when used to treat cells under anaerobic conditions, in which ROS are not generated (Liu and Imlay, 2013). This and accompanying data (Keren et al. 2013) have cast doubt and the role ROS play in antibiotic-induced bacterial cell death. Nevertheless, prolonged cellular perturbation through low-level antibiotic treatment might still lead to a condition in which ROS accumulate under aerobic conditions.

Several regulatory systems operate to manage cell envelope homeostasis during encounters with cell wall-targeting antimicrobial agents (Jordan et al. 2008). Other systems are induced by antibiotic treatment and respond by activating processes that remodel the cell envelope, detoxify or mobilize efflux systems designed to export the harmful agent (Cao et al. 2002b, Harwood et al. 1990; Jordan et al. 2008). Again, the two-component systems figure prominently in cell envelope stress response (Eldholm et al. 2010; Jordan et al. 2008; Nielsen et al. 1981; Suntharalingam et al. 2009). These consist of a membrane-bound histidine protein kinase bearing an extracellular sensory domain, and a response regulator activated by phosphotransfer from the sensory kinase. The stimulus that activates the system is derived from antibiotic treatment, although for most of these systems the exact signals that promote phosphorylation and phosphotransfer are not known. The cell envelope changes orchestrated by the various regulatory pathways functioning in the cell envelope stress response include a reversal in the negative charge of the cell wall through manipulation of teichoic acid and membrane phospholipid composition (Cao and Helmann 2004; Hyyrylainen et al. 2007; Jordan et al. 2008; Kristian et al. 2005). Genes specifying D-alanylation of teichoic acid and efflux transport systems for removal of toxic agents are activated.

Several of the TCSTSs that govern the cell envelope stress response are orthologs of the LiaRS (R: response regulator; S: sensor histidine kinase) system of *B. subtilis*. In *B. subtilis*, a third component, LiaF, is a membrane protein that exerts negative control over the LiaRS system (Eldholm et al. 2010; Jordan et al. 2008; Mascher et al. 2003; Nielsen et al. 2012; Suntharalingam et al. 2009; Wolf et al. 2010). Cell envelope stress induces other control pathways, namely the TCSTSs BceRS (Rietkotter et al. 2008; Ohki et al. 2003), PsdRS (Mascher et al. 2003) and YxdJK (Staron et al. 2011; Pietiainen et al. 2005), as well as the regulons governed by the ECF sigma subunits σ^M, σ^W, σ^X in *B. subtilis* (Cao and Helmann 2004; Cao et al. 2002a, b, Luo et al. 2010; Minnig et al. 2003, 2005). In the systems controlled by the aforementioned TCSTS, activation results in expression of ABC transporters that are encoded by genes that are cotranscribed with those specifying the TCSTS components. Recent studies have uncovered conserved antibiotic resistance modules within loci that specify the two-component signaling systems and the ABC transport apparatus that confer resistance (Gebhard, 2012; Dintner, et al. 2011; Staron, et al. 2011; Mascher, 2006). Interestingly, the transport systems play roles not only in antimicrobial peptide resistance, but also signal perception suggesting direct communication between transporter complex and the sensory kinase partner of the two-component pair. The sigma subunits, σ^M, σ^W, and σ^X in *B. subtilis* have distinct, but overlapping regulons (Luo et al. 2010; Mascher et al. 2007). Activation involves release from anti-sigma factors in response to signals derived from cell envelope stress (Heinrich and Wiegert 2009). For example, σ^W is controlled by RsiW, which is a membrane-bound anti-sigma that undergoes proteolytic destruction upon alkaline stress (Schobel et al. 2004; Zellmeier et al. 2006). The σ^W regulon is composed of approximately 60 genes that confer resistance to fosfomycin, a variety of bacteriocins and other antimicrobial peptides (Cao et al. 2002a, Huang et al. 1999; Pietiainen et al. 2005). The genes activated by the σ^X form of RNA polymerase encode products that catalyze D-alanylation of TA and synthesis of phosphatidylethanolamine, which together reduce the negative charge of the cell wall, rendering cells more resistant to cationic antimicrobial peptides (Cao and Helmann 2004; Minnig et al. 2003). The σ^M regulon is vast, encompassing genes not only required to maintain cell envelope homeostasis, but also genes that function in the oxidative stress response and in DNA repair (Eiamphungporn and Helmann 2008; Jervis et al. 2007; Thackray and Moir 2003). The *yjbCspxA* operon is one of the targets of the σ^M- and σ^W- forms of RNA polymerase in *B. subtilis* (Eiamphungporn and Helmann 2008; Jervis et al. 2007; Thackray and Moir 2003).

Genes encoding the phage shock proteins are activated through a LiaRS-dependent mechanism in *B. subtilis*, although the role of the phage shock components in alleviating cell envelope stress is unclear at this time (Wolf et al. 2010). However, the LiaRS systems have more responsibilities in other members of the *Firmicutes*, particularly those that lack multiple ECF sigma factors. For example, the LiaRS system of *Streptococcus mutans* is required to activate genes encoding products that function in cell wall biosynthesis, extracellular proteases and chaperones, and transcription factors, including the paralogous SpxA1 (Suntharalingam et al. 2009). Orthologous forms of *spx* are expressed in response to cell wall stress (see below), and at least one of these participates directly in cell wall remodeling (Veiga et al. 2007). *B. subtilis* Spx, when overproduced, stimulates the transcription of *yocH* (Nakano et al. 2003; Zuber et al. 2011), which encodes an extracellular peptidoglycan hydrolase required for optimal growth (Shah and Dworkin 2010).

Oxidative Stress Response

Aerobic metabolism is a preferred strategy for energy generation among many bacterial taxa, including members of the *Firmicutes*, but high concentrations of oxygen manifest toxic effects upon reaction with cellular components (Imlay 2003). Oxygen reaction with dihydroflavin cofactors results in sequential reductions that generate superoxide anion and hydrogen peroxide (Messner and Imlay 1999). These can react with a variety of compounds to heightened oxygen's toxic effects. Superoxide reacts with exposed iron–sulfur cofactors, releasing reduced iron that can react with peroxide to generate damage-inducing hydroxyl radical (Imlay 2006). A major toxic effect of reactive oxygen species is DNA damage resulting from reduced Fe-mediated hydroxyl radical formation (Imlay and Linn 1988). Enzymes bearing mononuclear ferrous iron within their active sites are susceptible to peroxide attack, which results in oxidation of coordinating Cys thiols (Anjem and Imlay 2012). ROS react with protein and low molecular weight thiols to upset the redox balance within the cytoplasm. Proteins and lipids are also targets of oxidative damage during cellular exposure to ROS.

The response to ROS involves the mobilization of genes encoding products that function to detoxify toxic oxidants and to repair the damage they cause (Storz and Spiro 2011). These include ROS-scavenging enzymes and low molecular weight thiols that can participate in detoxification by quenching ROS. Among the ROS-scavenging enzymes are the superoxide dismutases that convert superoxide to H_2O_2, which is a substrate for numerous other scavenging enzymes, such as catalases (mono and bifunctional) and several classes of peroxidases, including thiol peroxidases, peroxiredoxins, and cytochrome c peroxidases (Mishra and Imlay 2012). Organic hydroperoxides are substrates for alkyl hydroperoxide reductases, one being OhrA, a peroxiredoxin of *B. subtilis* and member of the OsmC/Ohr family (Fuangthong et al. 2001), but also found in a variety of Firmicutes members. The *Firmicutes*, especially members of *Bacillales*, produce a specialized low molecular weight thiol, bacillithiol (Helmann 2011), that is believed to function in processes normally associated with the role played by glutathione in many organisms. Recent studies have implicated these compounds in reactions that lead to modification and removal of other toxic agents (Chi et al. 2011; Gaballa et al. 2010; Newton et al. 2009), particularly thiol-reactive substances. One such study utilized thiol-reactive bimane derivatives, which showed that modified bacillithiol in *S. aureus* is processed to the mercpaturic acid, which is exported from the cytoplasm (Newton et al. 2012). Bacillithiol-producing species possess a conserved biosynthetic apparatus that bears some resemblance to the enzymes that catalyze the synthesis of mycothiol, a low molecular weight redox buffer, in members of the *Actinomycetales* (Gaballa et al. 2010; Newton et al. 2011). Both systems share a glycosyltransferase (BshA) in bacilli that links N-acetyl glucosamine to malic acid in bacillithiol synthesis (N-acetyl glucosamine to myo-inositol-1-phosphate in mycothiol synthesis) and a hydrolase (BshB) that removes the acetyl group to yield the glucosamine derivative. A Cys-adding enzyme establishes an amide link between the deacetylated glucosamine group and cysteine. Bacillithiol (BSH) is not only a redox thiol buffer, but also a metal chelator due to the malic acid group linked to glucosamine-Cys. Two genes encoding hydrolases (*bshB1* and *bshB2*) exist in *B. subtilis*; one of them, *bshB2*, is under the transcriptional control of Spx (Chi et al. 2011; Gaballa et al. 2010). The Cys-adding enzyme, BshC, is encoded by a gene (*yllA*) that is also activated by Spx (Chi et al. 2011). Based on ChIP analysis, *bshA* and *bshB1* transcription is partially Spx-dependent (Rochat et al. 2012).

ROS could potentially result in formation of BSSB, in which two BSH molecules become linked by a disulfide bond. For this there exists a putative BSH disulfide reductase encoded by the *B. subtilis ypdA* gene (Gaballa et al. 2010), another member of the Spx regulon (Nakano et al. 2003a), an observation that was supported by recent ChIP and expression data (Rochat et al. 2012). Additionally, there exists a bacillithiol-S-transferase that operates in a manner analogous to glutathione-S-transferase by catalyzing the bacillithiolation of toxic substrates targeted for elimination (Newton et al. 2011). Thus, evidence indicates that there exists a system to process thiol-reactive toxins and eliminate them through reaction with bacillithiol.

The genes encoding the ROS-scavenging enzymes and enzymes that catalyze bacillithiol biosynthesis are controlled by several regulatory factors, primarily at the level of transcription initiation. The PerR transcriptional repressor controls genes that function in the oxidative stress response and is conserved among members of the *Bacillales*, as well as in certain Gram-negative species such as *Campylobacter* (Fuangthong et al. 2002; Horsburgh et al. 2001; van Vliet et al. 1999). Genes controlled by PerR in *B. subtilis* include *katA* (heme catalase), *mrgA* (iron storage protein), *ahpCF* (peroxidases), *hemAXCDBL* (heme biosynthesis), and *fur* (ferric uptake regulator). PerR, along with another repressor, YodB, also negative controls the utilization of promoter P3 of the *spxA* gene (Leelakriangsak et al. 2007). The derepression of the PerR regulon results in upregulation of ROS-scavenging enzyme production and reduced intracellular iron, presumably to minimize damaging reactions between reduced iron and ROS. The secondary effect of this is reduced growth due to iron deficiency, not only due to Fur repressor accumulation, but heme-Fe titration by the high concentration of catalase generated by PerR repressor inactivation (Faulkner et al. 2012).

A number of genes functioning in elimination of ROS, in the biosynthesis of low molecular weight redox buffers, and in alleviating oxidative damage are activated through an Spx-dependent mechanism of transcriptional control. As *spx* is also a member of the PerR regulon, the peroxide-dependent inactivation of PerR elicits a secondary upregulation of genes we would normally associate with ROS damage repair. Further discussion of the Spx regulon and links to other pathways of stress response control is presented below.

General Stress Response Through Activation of σ^B and Involvement of Spx Paralogs

Genes encoding orthologs of SpxA are induced by conditions that activate the general stress response in members of *Firmicutes*. Inducing conditions are often sudden, dramatic changes in the environment that are lethal to microorganisms. Bacteria in the soil and rhizosphere are faced with extremes in temperature, nutrient/oxygen availability, and extremes in environmental water content resulting from rainfall and drought conditions (Fredrickson et al. 2008; Schimel et al. 2007). Such changes in the environment impart physical and chemical stress upon bacteria, resulting in reduced capability to generate energy, in the accumulation of ROS, and in elevated concentrations of denatured or damaged macromolecules. Fundamental changes to the gene expression machinery constitute part of a global stress response that is mobilized under harsh environmental conditions (Hengge 2011; Price 2011). One such change in *E. coli* involves RNA polymerase holoenzyme, which will undergo a shift in subunit composition resulting from the induction of *rpoS*, the gene encoding the general stress sigma subunit, σ^S (Hengge-Aronis 1993).

The response to environmental and energy stress in *Firmicutes* is complex and reflects the tasks undertaken to not only detoxify the afflicting agent, but to also cope with the damage it has caused. Some have distinguished the two tasks as primary and secondary stress responses, respectively (Mols and Abee 2011). When the toxic agent is first encountered, there is an induction of genes whose products might be required to carry out detoxification or immediate protection, while the damage caused by toxic agents, such as protein denaturation or oxidation (sulfenic acid and sulfenamide formation, carbonylation) might result in the induction of genes normally associated with the heat shock or oxidative stress responses. The effects of heat shock, osmotic stress, as well as the secondary effects of antimicrobial agents, including toxic oxidants, induce what has come to be called the general stress response (Petersohn et al. 2001; Price 2002).

The *spx* gene participates in the secondary response to environmental or energy stress, as it is a member of the σ^B regulon that is activated as part of the general stress response (Petersohn et al. 2001, Fig. 1). The σ^B system of control is conserved in many members of the *Firmicutes*, where it plays a role somewhat analogous to σ^S, although the means by which σ^B is controlled is quite different than that which operates to regulate σ^S. The *sigB* operon of *B. subtilis* contains the gene encoding σ^B and the genes *rsbW* and *rsbV*, which encode the anti-σ^B and inhibitor of anti-σ^B (anti–anti-σ^B). The operon is positively autoregulated, as it bears a promoter that is utilized by the σ^B-form of RNA polymerase holoenzyme. The organization of the *sigB* operon is shared with those of *L. monocytogenes* and *S. aureus*. Features of σ^B control are shared with the regulation of the sporulation-specific sigma factor σ^F (Duncan et al. 1994; Kang et al. 1998). Both σ^B and σ^F are negatively controlled through direct contact with the anti-sigma protein. Thus, RsbV inhibits RsbW, allowing σ^B to engage RNA polymerase, but RsbW-catalyzed phosphorylation reverses this inhibition. The anti-sigma possesses Ser/Thr kinase activity that targets the anti-sigma inhibitor, RsbV. The phosphorylation state of RsbV is controlled by phosphatases RsbU and RsbP. Thus, when the RsbU and RsbP are active, RsbV inhibitory activity is stimulated, resulting in release of σ^B from RsbW inhibition. Regulatory inputs are generated from energy depletion and environmental stress, and these input signals target RsbV activity.

The signal transduction systems that are thought to directly receive stress-related stimuli constitute two regulatory branches that converge on the activity of RsbV. Energy depletion triggers the RsbP-RsbQ system, where RsbP is proposed to be a signal receiver and bears a PAS domain, while RsbQ is a hydrolase that is thought to process the signal that affects RsbP phosphatase activity. Antimicrobial agents that impact ATP levels affect σ^B activity through the RsbPQ system of signal processing (Brody et al. 2001; Vijay et al. 2000). The second branch serves to activate the σ^B regulon in response to environmental stresses involving toxic agents such as high salt concentration, ethanol, heat, extremes of pH, and osmotic stress. A large protein complex, the stressosome (Marles-Wright et al. 2008), is believed to receive signals derived from encounters with toxic agents. This branch of the general stress response targets the phosphatase RsbU through the activator RsbT. RsbT is sequestered by the stressosomal proteins RsbR and RsbS (Akbar et al. 1997, 2001). A stress-induced signal, the exact nature of which is unknown, triggers the kinase activity of RbsT, leading to phosphorylation and inactivation of

RsbS and release of RbsT from the stressosome. Free RbsT stimulates RsbU phosphatase activity, allowing dephosphorylated RsbV to inhibit the anti-sigma factor, RsbW. The stressosomal proteins are the receptors of the stress-induced signal leading to elevated σ^B activity. Details of the stressosomal sensory mechanism are still unclear, however. The focus of research into the sensory function of the stressosome has been the RsbR paralogs. The RsbR proteins of the stressosome possess a conserved C-terminal STAS (sulfate transporter and anti-sigma factor antagonist) domain and an N-terminal nonheme globin-like domain that varies in structure and is thought to serve a sensory function (Murray et al. 2005). RsbR serves a positive and negative role in σ^B activation. The RsbR paralogs, along with RsbS, constitute the core of the stressosome and participate in sequestering RsbT, thus exerting negative control upon σ^B activity. They also serve a positive role by stimulating RsbT kinase activity upon stress induction, leading to RbsT release from the stressosome. The only RbsR paralog for which an activating signal has been determined is the YtvA protein, which senses blue light via its N-terminal LOV (light-oxygen-voltage) domain (Gaidenko et al. 2006). Notably, Spx is a positive regulator of *ytvA* transcription, indicating a role for Spx in σ^B control (Gaidenko et al. 2006). It is not clear at this time the regulatory logic that links Spx activation with the control of an RbsR stressosome component.

The complexity of the general stress-induced, σ^B signaling system of *B. subtilis* is not duplicated in all members of *Firmicutes. L. monocytogenes* bears the stressosomal proteins believed to receive stress-induced signals (Shin et al. 2010; Chaturongakul and Boor 2004; Ferreira et al. 2004), but produces only one phosphatase, RsbU, and no phosphatase dedicated to receive signals derived from reductions in energy production Thus energy stress induction of the σ^B is thought to reflect reduced ATP concentration that leads to a decline in RsbW kinase activity, rendering the anti-sigma protein more susceptible to anti–anti-sigma inhibition (Shin et al. 2010).

A different signaling system was uncovered in studies of *B. cereus*, where the RsbU phosphatase, rather than having N-terminal residues that mediate interaction with RsbT, possesses instead a response regulator receiver domain normally associated with TCSTS (de Been et al. 2010). The control of the phosphatase is conducted by the cognate histidine kinase RsbK, which is membrane bound, and bears an external CHASE (Cyclases/Histidine kinases Associated Sensory Extracellular) domain and a cytoplasmic GAF (cGMP-specific phosphodiesterases, adenylyl cyclases and FhlA) domain putatively involved in sensing salt and ethanol stress.

Extensive studies of the genes transcribed by the σ^B-form of RNA polymerase in *B. subtilis* have involved the generation of null mutations in 94 σ^B regulon members (Reder et al. 2012c). The stress sensitive phenotype of each member was analyzed, which employed a spectrum of stress conditions, including cold, salt, ethanol, paraquat, and peroxide. Nearly all mutations conferred sensitivity to multiple stresses, and there was a close correlation between loss of salt or cold resistance and sensitivity to paraquat, indicating that oxidative stress was a secondary consequence of stress exposure. Indeed, induction of the stress response mediated by σ^B by treatment with ethanol resulted in elevated resistance to paraquat, a superoxide cycling agent. Two members of the σ^B regulon that were tested in this study were the *spx* and *mgsR*, encoding paralogous proteins, both of which have been shown to activate transcription. *spx* transcriptional control is more complex, however (see below) as it is a member of multiple regulons governed by various stress-activated factors, whereas *mgsR* transcription is primarily governed by σ^B activity (Reder et al. 2008). Thus, *spx* serves as a node in the stress response network, while *mgsR* (modulator of general stress response), as its name implies, modulates the σ^B-activated general stress response.

The *spx* and *mgsR* genes are not the only members of the σ^B regulon that encode paralogous proteins. Several enzymes that catalyze the production of reducing equivalents exist in paralogous forms in *B. subtilis* and are encoded by σ^B regulon members. The mutational studies of Reder et al. (2012c) did not uncover evidence of functional redundancy, however, as all mutants, even those generated in paralogous genes, exhibited often multiple stress sensitive phenotypes. The observations reported in the study suggest that σ^B activation fosters an accumulative protective response involving multiple contributors participating in similar functions.

In this section, we have provided the foundation to address the premise presented in the introduction of this brief, that Spx is positioned as a node in the stress response network of Gram-positive bacteria. This assignment for Spx reflects the multiple convergent, stress-induced signals that activate *spx* expression, and the large number of genes specifying diverse functions that are affected, negatively and positively, by elevated Spx concentration and activity.

Expression of *spx* Is Affected by Multiple Stress Stimuli

Evidence of complex control over the expression of *spx* has accumulated over the years, and is based on data from several experimental approaches and from several sources (Zuber 2009). While there have been few studies that have examined the transcriptional control of the *spx* gene in Gram-positive bacteria, much of what is known about control of the *spx* gene comes from studies of *B. subtilis* involving transcriptomic and proteomic data, and studies of alternative forms of RNA polymerase (Cao et al. 2002a, Eiamphungporn and Helmann 2008; Helmann et al. 2001; Jervis et al. 2007; Petersohn et al. 2001). In *B. subtilis*, the *spx* gene resides in a dicistronic operon with the upstream *yjbC* gene, encoding a putative acetyltransferase. Multiple promoters recognized by the major σ^A form of RNA polymerase, and at least three alternative RNA polymerase forms control the *yjbCspxA* operon (Fig. 3).

Microarray transcriptome analysis provided evidence that *spx* was part of the general stress response, with elevated transcript levels upon phosphate limitation and heat shock (Antelmann et al. 2000; Petersohn et al. 2001). The induction of *spx* was attributable in part to σ^B activation, resulting in utilization of a σ^B promoter upstream of the *yjbC* gene. Primer extension analysis confirmed the presence of a start site utilized by σ^B as well as an upstream promoter recognized by σ^A. The *yjbC* gene

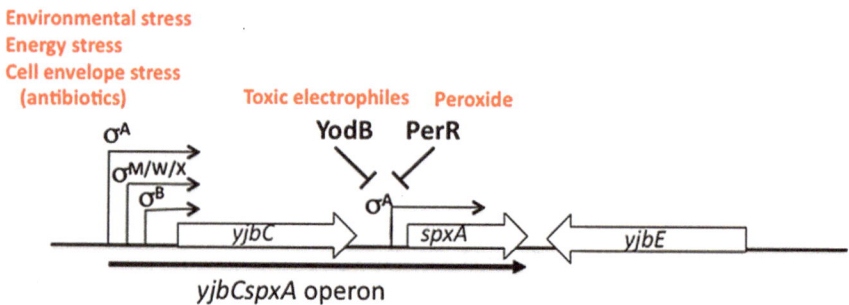

Fig. 3 The organization of the *spxA* locus of *Bacillus subtilis*. The *yjbCspxA* operon is indicated by the *black arrow* beneath the *white block arrows*. The diagram shows the locations of the *yjbC*, *spxA,* and *yjbE* coding sequences (*white box arrows*) in part of the B. subtilis chromosome. *Small black arrows* indicate the positions of promoters within the *yjbCspxA* operon. Transcription factors controlling *yjbCspxA* are indicated. *Flat-headed arrows* indicate negative control. Stress conditions are denoted by *red text*

was identified as a member of a regulon that is transcriptionally activated by the σ^M form of RNA polymerase in response to cell envelope stress (Eiamphungporn and Helmann 2008; Jervis et al. 2007). A thorough study of the σ^M regulon involving microarray, promoter consensus sequence search, and in vitro run-off transcriptome microarray analysis (ROMA) also revealed that *yjbCspxA* was transcriptionally activated by the σ^M-RNA polymerase form and under cell wall stress in a σ^M-dependent manner (Eiamphungporn and Helmann 2008). The σ^M-utilized promoter is also the target for other ECF sigma-bearing RNA polymerase forms, σ^W and σ^X, that are activated after treatment with cell wall-active antimicrobial agents (Cao et al. 2002a, Mascher et al. 2007; Pietiainen et al. 2005). The sigma subunits σ^W and σ^X show overlapping promoter sequence specificity, and show partially overlapping regulon composition (Mascher et al. 2007).

A paralog of SpxA in *B. subtilis* is the product of the *mgsR* gene, which is a member of the σ^B regulon (Reder et al. 2008). Unlike *spxA*, *mgsR* is transcribed from one promoter, which is recognized by the σ^B form of RNA polymerase (Reder et al. 2008, 2012d), and does not exhibit the complexity that characterizes *spxA* transcriptional control.

The organization of the *yjbCspxA* operon is conserved in members of the genus *Bacillus* that are closely related to *B. subtilis*, including *B. pumilus*, *B. atrophaeus*, *B. licheniformis*, and *B. amyloliquefaciens*. While other Bacilli, such as *B. megaterium* possess a *yjbC* gene, it is not cotranscribed with the *spx* gene. In most members of the *Bacillales*, including the *B. cereus/anthracis/thuringiensis* group, *spxA* appears to be monocistronic and a *yjbC* ortholog is absent. In *B. anthracis*, there are two genes encoding SpxA paralogs, *spxA1,* and *spxA2*. The *spxA1* gene is likely monocistronic, while *spxA2* is the first gene of a dicistronic operon that also bears a gene encoding a putative low molecular weight chaperone.

The orthologous *spx* genes of *S. mutans*, *S. pneumoniae*, and *L. lactis* are induced by envelope stress through the LiaFSR system that is conserved in *B. subtilis*. In *S. pneumoniae*, *spx* is activated by treatment with bacitracin, tunicamycin, and nisin, but not by vancomycin (Turlan et al. 2009), which induces the LiaSR system in *B. subtilis*. A number of antibiotics can induce the competent state, including aminoglycosides and fluoroquinolones, but antibiotics that target the cell wall do not induce competence (Prudhomme et al. 2006). While it is known that SpxA1 can inhibit competence (Turlan et al. 2009), it is not known if this is a result of cell envelope stress that might lead to elevated *spxA1* expression. The *spxA* gene of *S. mutans* is induced by bacitracin treatment in an LiaSR-dependent manner (Eldholm et al. 2010; Suntharalingam et al. 2009). The LiaSR orthologous system in *L. lactis*, CesSR, regulates the *spxB* gene, whose product activates the *oatA* gene in response to antimicrobial hydrolytic degradation of peptidoglycan (Veiga et al. 2007). The *oatA* gene encodes a peptidoglycan O-acetylase that modifies *L. lactis* peptidoglycan, rendering it resistant to antimicrobial hydrolytic activity. A paralog of *spxB*, *trmA* was the site of mutations that affect the secretion of lysostaphin in a genetic engineered derivative of *L. lactis*, suggesting that the paralog encoded by *trmA* might also function in control cell wall metabolism (Tan et al. 2008). The logic that links cell wall perturbation, induction of *spx* gene transcription and Spx function are clear in *L. lactis* as there is evidence for Spx-dependent changes within the arsenal of cell wall-active enzymes. Further evidence for a role of Spx in cell envelope dynamics comes from studies of the opportunistic pathogens *S. aureus* and *S. epidermidis*, in which Spx was shown to negatively control production of the polysaccharide intercellular adhesin (PIA), a prerequisite for biofilm development (Pamp et al. 2006; Renzoni et al. 2011; Wang et al. 2010). This is accomplished by modulating the expression of the *icaADBC* operon specifying the enzymes that catalyze β-1,6-linked glucosaminylglycan synthesis. Mutants with lesions in the *clpP* gene show reduced capability to produce biofilms, attributable to the resultant accumulation of Spx protein (Michel et al. 2006). The effects on *icaADBC* expression in *S. aureus* were explained as being due to the upregulation of the *icaR* gene, encoding a repressor of the capsule biosynthesis operon (Cue et al. 2012; Pamp et al. 2006). However, the Spx-dependent negative control of *icaADBC* in *S. epidermidis* did not appear to be connected to IcaR (Cue et al. 2012; Wang et al. 2010). How synthesis of Spx is regulated in *Staphylococcus* species is unclear at present, although there is evidence that the concentration of Spx is governed by regulated proteolysis involving ClpXP (Pamp et al. 2006).

What is less clear is the relationship between cell wall stress and the upregulation of *Bacillus spx* in terms of what might be accomplished by activating the *spx* regulon. A simple answer would involve citation of numerous reports that connect exposure to cell wall-active antibiotics with bacteriocidal production of ROS (Kohanski et al. 2010). Thus, the activation of σ^M in *B. subtilis* by encounters with cell wall-acting antimicrobial agents would result in accelerated transcription of *yjbCspxA*, leading to activation of a regulon in which most of the members function in alleviating oxidative stress that is a secondary consequence of antibiotic treatment. However, some of the genes activated by Spx may be more directly linked to cell envelope metabolism, as thioredoxin, the gene for which is a member of the Spx regulon, is known to function in electron delivery to the periplasmic space in Gram-negative bacteria (Ritz and Beckwith 2001). A more detailed look at the targets of Spx-activated transcription is presented in the next section.

Construction of transcriptional *lacZ* fusions to the *spx* gene uncovered a fourth promoter residing in the intergenic region between *yjbC* and *spx* of *B. subtilis* (Leelakriangsak et al. 2007, 2008; Leelakriangsak and Zuber 2007). In vitro transcription with purified RNA polymerase confirmed the existence of a promoter that was recognized by the σ^A form of RNA polymerase. The low transcriptional activity of the promoter in vivo was attributable to two genes encoding transcriptional repressors. PerR is the peroxide regulator involved in repression of catalase and peroxidases genes, and interacted with an element immediately downstream from the transcriptional start site of *spx*. The second repressor was YodB (Chi et al. 2010; Leelakriangsak et al. 2007, 2008), which is inactivated by reaction of an N-terminal thiol with an electrophile, such as an oxidized quinone. The logic that links PerR and YodB function, *spx* transcriptional induction and the function of the activated Spx regulon is clear, as peroxide and toxic electrophiles can be viewed as agents promoting disulfide stress.

Evidence from transcriptomic studies suggest that expression of the *spx* gene in *B. subtilis* is controlled differently compared to orthologous *spx* genes in other members of the *Bacillales*. Studies of nitric oxide-treated cultures of *B. subtilis* and *S. aureus* showed striking differences in proteomic changes (Hochgrafe et al. 2008). Nitric oxide (NO) exposure caused a marked increase in proteins encoded by genes of the Spx regulon in *B. subtilis*. However, Spx-controlled genes, such as *trxB*, in *S. aureus* were not upregulated by NO treatment. There was no induction of genes normally associated with the oxidative stress response in *S. aureus* when cultures were exposed to NO, which instead promoted the induction of genes functioning in anaerobic/fermentative metabolism (Hochgrafe et al. 2008). The result supports the view that orthologous forms of Spx may serve different roles in certain members of the *Bacillales*, as was shown in the case of SpxB in *L. lactis*.

Recently reported high-throughput transcriptomic analyses using massively parallel RNAseq technology have uncovered over a genome-wide scale the growth and stress conditions that affect gene transcript levels in *B. subtilis*. The study examined cellular transcript levels in cultures grown in or exposed to a total of 269 conditions (Nicolas et al. 2012). The data compiled include the conditions under which the highest transcript concentrations were observed for each gene. From the data, the conditions promoting the highest *spx* transcript levels were uncovered, and indicated that *spxA* transcript concentration was highest when cells were exhibiting multicellular behavior. Cells undergoing confluent colony growth, biofilm formation, or swarming showed the highest levels of *spx* transcript as did cells treated with ethanol. The observations suggest that accelerated *spx* transcription is associated with changes to the cell surface/envelope and cell density. The ECF sigma subunits σ^M, σ^W, and σ^X have been implicated in efficient biofilm formation, and σ^B also was observed to play a role in biofilm through transcription of the *yxaB* gene (Nagorska et al. 2008; Nicolas et al. 2012). Thus, heightened ECF sigma factor and σ^B activity might be a characteristic of cells that are participating in multicellular behavior, and could account for the elevated levels of *spx* transcript. The accelerated transcription implied by the RNAseq results does not seem to translate into heightened Spx regulon activity, however. The Spx-controlled *trxB* gene shows the highest transcript levels in diamide-, peroxide-, and ethanol-treated cells, according to the RNAseq data (Nicolas et al. 2012). This might reflect the additional layers of control acting upon Spx concentration and activity, including redox and proteolytic regulation.

The paralog of *spx* in *B. subtilis*, *mgsR*, is a member of the σ^B regulon, but is transcribed by a single σ^B-utilized promoter. Analysis of ectopic expression of an *mgsR* construct and mutant derivatives in a strain bearing a deletion of *mgsR* at its native locus showed that *mgsR* transcription is under positive autoregulation (Reder et al. 2012d). Thus, *mgsR* expression is influenced by a number of stress conditions that activate σ^B, but is also stimulated by potential secondary effects, such as oxidative stress that elevate MgsR activity through its dithiol/disulfide switch.

Several transcriptomic experiments of *B. anthracis* RNA employing microarray and RNAseq technology have been undertaken and have shed light on the conditions under which the paralogous *spxA* genes, *spxA1* and *spxA2*, are transcriptionally active (Bergman et al. 2006, 2007; Passalacqua et al. 2007, 2012). In nearly all studies, *spxA1* (BA1200) was shown to have a consistently higher level of expression than *spxA2* (BA3456) and expression occurred at different points in the growth curve and under different conditions. Microarray studies of the *B. anthracis* transcriptome over the growth curve uncovered five temporal waves of transcription distinguishable by their transcriptomic patterns (Bergman et al. 2006). The *spxA1* gene transcripts were observed in wave 1, which is soon after spore germination, while *spxA2* was expressed late in the sporulation process (wave 5). Neither protein was found in the spore proteome of *B. anthracis*, however (Liu et al. 2004). Elevated expression of *spxA1* after germination might reflect a response of the outgrowing cell to mobilize a defense against potential oxidative damage that accompanies the sudden exposure of the germinated spore to an influx of water and oxygen. Support of this notion comes from transcriptomic studies of *B. anthracis* cultures under stress conditions, which generated data showing induction of *spxA1* by treatment with paraquat and peroxide (Pohl et al. 2011). A recent RNAseq transcriptomic study uncovered elevated *spxA1* transcript levels upon ethanol treatment, potentially linking induction to the general stress response system (Passalacqua et al. 2012). Transcriptional profiling was conducted that sought to examine changes in the *B. anthracis* transcriptome during macrophage infection (Bergman et al. 2007). The *spxA2* gene was 12-fold upregulated during late outgrowth stage in the macrophage compared to expression in cultured cells, while *spxA1* is induced 3.3-fold. The induction of *spxA2* in the infected macrophage is in contrast to the generally low expression observed in culture (S. Barendt and P. Z., unpublished). The *spxA1* gene resides in a genomic context that resembles that of other Gram-positive bacteria, while the locus occupied by *spxA2* is unique, and shows no synteny with members of *Bacillales* outside of the *B. cereus/anthracis/thuringiensis* group. Both genes are preceded by a large intergenic sequence of 514 bp within which no gene or coding sequence has been assigned. How these regions function in transcriptional control of the *spxA*1 and *spxA2* has yet to be determined. The transcriptomic data indicate that the control of *spxA1* is in line with what has been observed for *spxA* in *B. subtilis*. The conditions under which *spxA2* becomes transcriptionally active requires further exploration.

Post-Transcriptional Control of Spx: Redox-Dependent Activation

Several structural features of the Spx protein have been linked to its role in transcriptional repression and activation (Nakano et al. 2010; Newberry et al. 2005). The protein consists of two domains. Domain 1, the redox control domain, bears the N-terminal redox center and the C-terminal coil required for proteolytic regulation. The coil emanates from a 4-stranded β sheet, also in domain 1. Domain 2 is formed from the central region of the protein that contacts the C-terminal domain of the RNA polymerase α subunit. It also contains α4, a short helix that data suggest functions in target DNA interaction (Nakano et al. 2010). A residue substitution G52R in domain 2 disrupts interaction with αCTD and renders Spx defective in both transcriptional repression and activation. Changing the reactive C10 to Ala results in loss of transcriptional activation (Nakano et al. 2005) and reduced, but not complete loss of repression (Zhang et al. 2006). This mutation weakens, but does not eliminate binding of Spx to the α subunit (A. Lin and P. Zuber, unpublished). These results were confirmed in vivo using *lacZ* reporter constructs in *B. subtilis*, and in vitro using purified RNA polymerase, Spx or mutant derivatives, and linear promoter DNA templates in run-off in transcription reactions.

The Spx protein is a structural homolog of ArsC (arsenate reductase), but no enzymatic activity has been detected in purified Spx protein preparations. There is weak primary structure similarity between ArsC and Spx, but certain conserved residues suggest functional roles in Spx activity. A distinguishing feature of Spx is the CXXC motif of domain 1 near the N-terminus, which contains the active site Cys (Cys10 in Spx) that is shared with ArsC. In ArsC, the conserved Cys is covalently linked to arsenate to form the reaction intermediate, monohydroxy-thioarsenite, in arsenate reduction (Martin et al. 2001). ArsC lacks the resolving Cys (Cys13) in Spx, which together with Cys10 forms the CXXC redox disulfide center of Spx. A triad of Arg residues (R60, R94, R104) with side chains in close proximity to the active thiol of arsenate reductase function in thiol reactivity and efficient release of arsenite (Martin et al. 2001; Shi et al. 2003). In *E. coli*, glutathione and glutaredoxin function in ArsC-catalyzed arsenate reduction (Oden et al. 1994; Shi et al. 1999).

The R92 of Spx (Fig. 4), corresponding to the R94 residue of ArsC, is in position to make electrostatic contact with C10 thiolate. Oxidation to the C10-C13 disulfide results in a shift in R92 side chain position away from the redox center and toward the α helix designated α4 (Nakano et al. 2010). Structural studies of a

Fig. 4 Structure of Spx and alpha CTD adopted from Lamour et al. (2009). Shown are residue side chains of R60 (*green*), R91, and R92 (*orange*). Location of Cys 10 (C10) is indicated

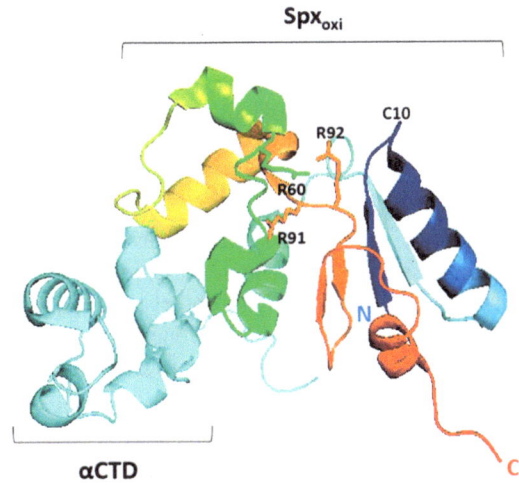

C10S mutant showed that the structure of α4 changes during reduction, unfolding into a coiled structure. The result suggested that redox control involved a repositioning of residue side chains within α4 (residues R60 to N68), which could affect Spx-activated transcription either by changing affinity to RNA polymerase or a target promoter sequence. Experiments employing electrophoretic mobility shift analysis (EMSA) using Spx or mutant derivatives, purified αCTD, and target promoter DNA showed that αCTD formed a shifted complex on the EMSA gel, but addition of Spx formed a supershifted complex (Nakano et al. 2010). A reaction containing SpxG52R and αCTD showed only the αCTD/promoter complex, while addition of SpxR60E to the αCTD/promoter reaction resulted in no shifted complex. The results confirmed that G52R conferred a defect in RNA polymerase interaction. The SpxR60E results suggested that the mutant Spx protein could interact with αCTD, but that it prevents αCTD from engaging promoter DNA, perhaps by masking part of its DNA-binding surface. This result further suggested that Spx, while masking part of the αCTD DNA-binding surface, most likely provides part of the DNA-binding function to the Spx/αCTD complex for specifically engaging the cis-acting element of Spx-controlled genes, a function that is lost when R60 is changed to glutamate (Nakano et al. 2010). The nature of the cis-acting element recognized by Spx/RNA polymerase complex is discussed below.

Recent epistasis studies involving the construction and analysis of double residue substitutions implicated R92 in Spx as functioning in redox control. Using in vivo reporter *trxB-lacZ* construct, the R92A substitution was found to have no effect on the *lacZ* expression when introduced into the Spx C10A mutant derivative. However the Spx mutant bearing R92A and the α4 helix residue substitution K62E showed a *trxB-lacZ* activity that was lower that that conferred by either substitution. Thus the R92 mutant phenotype is only observed when the protein is able to undergo oxidation to the disulfide state (A. A. Lin and P. Z. unpublished). The change in R92 side chain position caused by the redox state of the protein may influence the structure of α4 and positioning of the residues within so as to affect the DNA-binding surface. An interesting question is whether the promoter specificity of Spx paralogs is dictated by the helix α4 residues. Further investigation of Spx orthologs and attempts at constructing chimeric derivatives might provide data for answering this question.

The R104 of ArsC necessary for arsenate interaction is missing in Spx/MgsR and the YffB proteins of *Brucella* and *Pseudomonas*. In its place is a conserved Gly residue in a position after the last β strand of the C-terminal end. The absence of the R side chain is thought to provide a binding pocket for a substrate other than arsenate, and YffB was observed to bind glutathione (Teplyakov et al. 2004). However, evidence that glutathione participates in arsenate reduction as catalyzed by ArsC (Oden et al. 1994; Shi et al. 1999) suggests that the Arg at position 104 does not interfere with interaction between low MW biothiols and ArsC-family proteins. Nevertheless, the question of whether or not SpxA can interact with a low molecular weight biothiol as part of the mechanism of SpxA redox turnover needs to be addressed.

Recent studies of the *B. subtilis* MgsR have shown that redox control also affects the Spx paralog, as it bears the N-terminal CXXC and mass spectrometric analysis with thiol-reactive agents, along with mutational studies demonstrated that the motif serves as redox disulfide center (Reder et al. 2012d). Ethanol stress also caused an intermediate level of MgsR oxidation, providing evidence that ethanol induces oxidative stress as a secondary effect. As with Spx, MgsR in the oxidized disulfide state is active in stimulating transcription of genes under its control.

Post-Transcriptional Control of Spx: Regulated Proteolysis

Western blot analysis of Spx after thiol stress induced by diamide treatment of *B. subtilis* cells resulted in a transient increase of Spx protein concentration, with protein levels increasing after 10 min and subsiding after 60 min post-treatment (Nakano et al. 2003a). The same pattern was observed when the *spx* gene was expressed from an IPTG-inducible promoter, indicating that the control of Spx concentration was exerted at the post-transcriptional level. Accumulating evidence pointed to the ATP-dependent protease, ClpXP, as an important player in post-transcriptionally controlling Spx concentration (Nakano et al. 2001, 2002, 2003b). This was supported by the aforementioned phenotype of the *clpX* mutant and the in vitro reconstruction of Spx proteolysis catalyzed by ClpXP. Rapid degradation of Spx by ClpCP was demonstrated in vitro when one of the adaptor proteins, MecA or YpbH, was present (Nakano et al. 2002). Mutations in *clpC* or *mecA* had little effect on the level of Spx in vivo, however, suggesting that ClpXP proteolysis is masking the Spx-targeted activity of ClpCP. That SpxA is barely detectable in cells during unperturbed growth in culture is because ClpXP catalyzes its degradation.

As with other ClpXP substrates, the C-terminal end of *B. subtilis* SpxA was observed to be necessary for Spx to serve as a substrate. When the C-terminal end of Spx was changed from LAN to LDD, Spx was protease resistant and could be produced in high concentration in *B. subtilis* cells (Nakano et al. 2003b). Both SpxA1 and SpxA2 of *B. anthracis* could also be produced to high concentration by replacing the AN residues at the C-terminus with DD (S. Barendt, M. M. Nakano, and P. Zuber, unpublished). MgsR appears to be a substrate for ClpXP proteolysis based on the *B. subtilis clpX* mutant phenotype, but does not share C-terminal residue sequence homology with Spx (Reder et al. 2012d). The C-terminal end of SpxA is highly conserved among orthologs of *Bacillales*, but the *Staphylococcus* sp. bear a VD at the end instead of AN, and the L at the third position from the end is often replaced with M. It is not known how these changes might affect proteolytic control of Spx if it operates in other members of *Bacillales*.

Adaptor proteins are substrate recognition factors that bind substrate proteins and tether them to the protease, thus accelerating the rate of proteolysis (Kirstein et al. 2009). For example, MecA and YpbH function as adaptors, or substrate recognition factors, for ClpCP-catalyzed proteolysis in *B. subtilis* (Persuh et al. 2002; Schlothauer et al. 2003). The most studied adaptor of ClpXP is the SspB protein of *E. coli*, which recognizes the SsrA tag appended to products of interrupted translation by tmRNA (Hersch et al. 2004). SspB makes contact with both substrate and ClpX to facilitate delivery of the substrate for proteolysis. The discovery of a mutation in the *yjbH* gene that results in elevated SpxA concentration without affecting *spx* gene transcription (Larsson et al. 2007) suggested that the *yjbH* product might serve an adaptor/chaperone function. The mutant phenotype was similar to that of a *clpX* mutant, with high levels of Spx and defects in growth. The *yjbH* gene was observed to be a member of the Spx regulon and induced by NO treatment (Larsson et al. 2007; Rogstam et al. 2007). The *yjbH* gene is often linked to *spx* and its product is conserved in *Bacillales*, but absent in Streptococci. The product shows some similarity to disulfide isomerase DsbA, but no activity related to Cys thiol chemistry has been investigated.

YjbH protein from *B. subtilis* and *Geobacillus thermodenitrificans* was purified and shown to accelerate Spx degradation in a reaction containing ClpXP and ATP (Garg et al. 2009). YjbH from *G. thermodenitrificans* exhibited Spx-binding activity in an affinity pull-down reaction, but unlike SspB, no evidence was obtained that it could bind ClpX, whether in the presence or absence of ATP or a nonhydrolyzable analog (Chan et al. 2012). Small quantities of YjbH can mediate rapid ClpXP-catalyzed degradation of Spx, even at a 30:1 substrate to YjbH ratio, again unlike SspB, which was utilized in ClpXP proteolysis reactions with 0.5–1.0 molar equivalents of substrate (Chan et al. 2012).

While YjbH dramatically accelerates Spx degradation, there is no evidence yet that it functions in a sensory capacity during the thiol stress response. The N-terminus of the *B. subtilis* YjbH bears a Zn-binding domain that is sensitive to peroxide, leading us to propose that it might serve as a sensory module that affects YjbH activity in response to oxidation (Garg et al. 2009). However mutations in the domain that substituted Cys and His residues with Ala showed no loss of *yjbH* complementing activity in vivo or responsiveness to diamide treatment

(Chan et al. 2012; Engman et al. 2012). Substitution of all the Cys residues in the *S. aureus* YjbH protein resulted in no loss of activity, with *yjbH* complementing activity retained in *B. subtilis* (Engman et al. 2012). A possible mechanism for controlling YjbH activity was uncovered by yeast two-hybrid analysis, which identified a small protein, YirB that bound to YjbH, and prevented YjbH-mediated Spx degradation by ClpXP in vitro (Kommineni et al. 2011). Overexpression of *yirB* from an IPTG-inducible promoter resulted in an increase in Spx concentration, supporting the view that YirB competitively inhibits YjbH-Spx interaction. Affinity-tagged YirB captured YjbH in vivo and a YirB/YjbH complex could be recovered (Kommineni et al. 2011). However, a deletion of the *yirB* gene showed no phenotype, and expression of a *yirB-lacZ* fusion appeared to be constitutive and unaffected by oxidative stress. A role for YirB in control of Spx, if any, has yet to be determined.

Studies of ClpXP-catalyzed Spx degradation suggested that the ATPase subunit, ClpX might play a sensory role in controlling Spx levels in response to thiol stress (Zhang and Zuber 2007). The N-terminus of ClpX is a Zn-binding domain in which the zinc atom is coordinated by Cys residues that might be susceptible to oxidative Zn release. Mutations conferring amino acid substitutions at the Cys positions rendered ClpX unable to form an active complex with ClpP for the degradation of Spx. Furthermore, diamide addition to a ClpXP proteolytic reaction rendered Spx, and a mutant derivative bearing a C10A substitution in the redox center, resistant to proteolysis. The result confirmed that disulfide formation in Spx was not the reason for diamide-dependent Spx resistance to ClpXP-catalyzed proteolysis. In contrast, ClpCP/MecA-catalyzed Spx degradation was insensitive to concentrations of diamide that inhibited ClpXP. Thus, a model of proteolytic control can be proposed in which ClpX becomes oxidatively inactivated, allowing the accumulation of Spx. However, YjbH, which is likely to also increase in concentration, since transcription of its gene is activated by Spx, can conceivably titrate Spx and reduce its capacity to further stimulate its regulon. Perhaps, YirB binds to YjbH during oxidative activation of Spx, allowing the accumulating Spx protein to gain unhindered access to RNA polymerase. One could imagine that YirB, like Spx, is a substrate for ClpXP (Fig. 5). While it might be produced constitutively, it would only accumulate during oxidative stress, when it would effectively inhibit YjbH contact with Spx. Investigations of the in vivo oxidation state of ClpX and the susceptibility of YirB to ClpXP proteolysis have yet to be undertaken.

Proteolytic control of Spx by ClpXP might be conserved in other Firmicutes. Mutations in *S. mutans clpX* and *clpP* result in increased acid resistance and long-term survivability, but slower growth rate (Kajfasz et al. 2009, 2010). These mutations could be partially suppressed by deleting one of the two *spx* paralogous genes. Western analysis showed that both proteins, SpxA and SpxB, accumulated in the *clpX* mutant. No *yjbH* gene exists in *S. mutans* or other Streptococci, and the ClpXP-catalyzed degradation of Spx has yet to be demonstrated. Whether or not efficient Spx degradation in Streptococci requires an adaptor protein is also not known at this time.

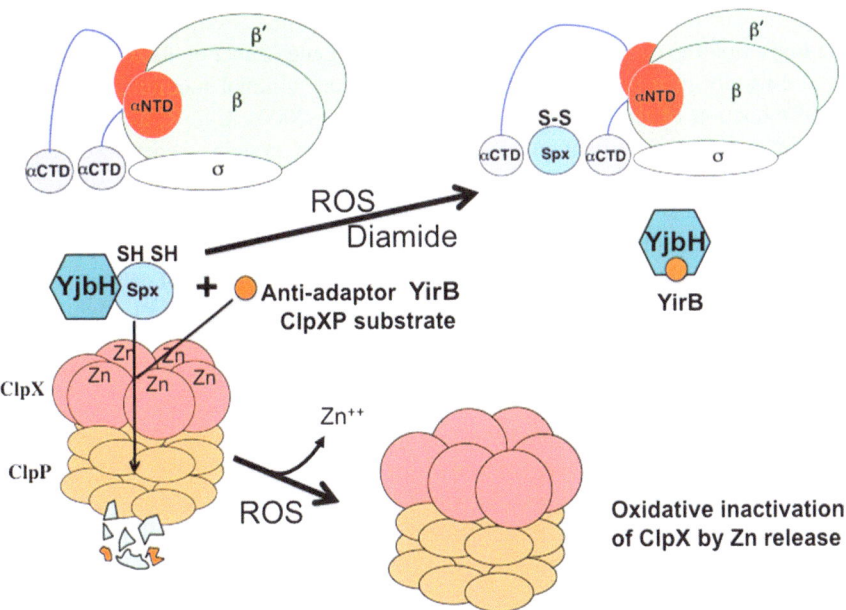

Fig. 5 Post-translational activation of Spx. Proposed activation mechanisms of Spx involving oxidation of Spx to disulfide state and increase in Spx concentration resulting from oxidative inactivation of ClpX. Under pre-stress conditions Spx is degraded by YjbH-mediated ClpXP-catalyzed mechanism. The anti-adaptor YirB, which inhibits YjbH, is also a ClpXP substrate and is degraded. Upon oxidative stress resulting from encounter with ROS or diamide treatment, Spx is oxidized and has higher affinity for RNA polymerase. ClpX is oxidatively inactivated by Zn release and YirB peptide accumulates to inhibit YjbH

Targets of Spx-Dependent Transcriptional Activation

Three studies have been reported that were undertaken to define the Spx regulon. The first was conducted to uncover the genes that were induced or repressed when Spx interacted with RNA polymerase (Nakano et al. 2003a). This employed microarray analysis of RNA isolated from a strain of *B. subtilis* that produced a protease resistance form of Spx (Spx[DD]). RNA was also purified from a strain producing Spx[DD], but which also bore the *rpoAY263C* mutation conferring a defect in Spx interaction with the α subunit of RNA polymerase. Many of the genes activated were also found in an independent study to be induced by treatment with diamide (Leichert et al. 2003). Thus genes encoding thioredoxin, thioredoxin reductase, methionine sulfoxide reductase, and genes required for the synthesis and uptake of cysteine were induced when Spx interacted with RNA polymerase. In all, 132 genes were identified that showed 3-fold or more induction by Spx-RNA polymerase interaction. Validation through primer extension of in vivo transcripts and reconstitution of Spx-activated transcription in vitro was achieved for several of the genes that were identified in the microarray hybridization analysis (Nakano et al. 2003a, 2005).

A second microarray experiment was reported which examined the change in the transcriptome caused by a deletion of the *spx* gene (Zuber et al. 2011). Deletion of *spx* results in a partial cysteine auxotrophy on minimal medium, hence there is an induction of the CymR regulon (Even et al. 2006). CymR, a DNA-binding protein and Rrf2 family member, in complex with cysteine synthetase (CysK), represses genes whose products function in Cys biosynthesis from sulfide and Met, and products that function in Cys uptake (Tanous et al. 2008). Deletion of the *spx* gene also results in derepression of genes that are members of the Fur (ferric uptake regulator) regulon, including genes encoding enzymes that catalyze sidero-phore biosynthesis and iron uptake (Ollinger et al. 2006). This effect might be due to reduced expression of *ytpQ*, a gene of the Spx regulon and encoding a product of unknown function (Zuber et al. 2011). Deletion of *ytpQ* also results in derepression of the Fur regulon genes. It is unknown how *ytpQ* affects control of iron homeosta-sis/metabolism. Genes controlled by regulators that function in sporulation initia-tion were also derepressed by the *spx* mutation. It is not known if the partial Cys auxotrophy of the *spx* mutant is responsible for some of the changes in transcript levels observed in the reported transcriptomic analysis of the *spx* null mutant.

Recently, a ChIP-Chip analysis using tiling genomic arrays was reported, which involved precipitation of an affinity-tagged derivative of *B. subtilis* Spx after diamide treatment (Rochat et al. 2012). The crosslinked complexes recovered contained RNA polymerase and the 144 promoter regions controlling 257 genes, many of which were genes uncovered by previous microarray experiments as being activated or repressed. Many of the same classes of genes previously identified as being controlled by Spx were uncovered, including those required for cysteine biosynthesis and for thiol homeostasis. Additionally, genes required for Cys uptake and electrophile detoxifica-tion were identified. Interestingly, some from the latter class of genes are also subject to control by other trans-acting positive and negative regulatory factors. Some of the operons positively controlled by Spx are also activated by HypR (Palm et al. 2012), which is responsive to NaOCl and diamide treatment.

The ChIP-Chip analysis and previous transcriptomic work provided evidence linking Spx-activated transcription with Spx proteolytic control. Transcript levels of *clpX* and *yjbH* increased after induction of Spx^{DD} production or by Spx-dependent induction after diamide treatment (Leichert et al. 2003; Nakano et al. 2003a, Rochat et al. 2012). As suggested previously (Larsson et al. 2007; Rochat et al. 2012), a mechanism of negative autoregulation is established by elevating *clpX* and *yjbH* expression through Spx-mediated control, which ensures the rapid removal of Spx when stress conditions subside. A negative autoregulatory circuit was described in the case of the actinomycetes σ^R RNA polymerase subunit, which activates gene expression in response to thiol oxidative stress (Kim et al. 2009). Among the mem-bers of the σ^R regulon, which includes those encoding thioredoxin, thioredoxin reductase, and other genes also induced by Spx in *B. subtilis*, are the genes speci-fying the two isoforms of ClpP. The *clpP* genes reside in a dicistronic operon with the *clpX* gene positioned downstream of the *clpP* operon. As with the autoregulatory circuit of Spx proteolytic control, the system in actinomycetes has been proposed to ensure quick inhibition of σ^R when thiol homeostasis is restored.

The regulatory sequences that were recovered by ChIP were aligned in an attempt to generate a candidate cis-acting regulatory sequence for optimal Spx-activated transcription (Rochat et al. 2012). However, deriving a consensus sequence for specific Spx/RNA polymerase-promoter interaction was deemed difficult based on the alignment that was analyzed. Nevertheless, a conserved GC at positions near −43, −44 with respect to the transcription start site, followed by an A-T-rich sequence upstream of the promoter −35 seemed to characterize the DNA captured by ChIP. This target sequence was supported by previous mutational analysis that provided evidence that the nucleotide positions identified by the alignments were also important for Spx/αCTD complex interaction, for activity in vivo, and transcription in vitro (Nakano et al. 2010). Such cis-acting elements utilized by Spx paralogs have not been uncovered thus far.

Transcriptomic analysis of the MgsR regulon uncovered a subset of the genes that are expressed when σ^B is active (Reder et al. 2008, 2012d). Although there is little overlap between the composition of the Spx and MgsR regulons, several of the MgsR-controlled genes are thought to specify functions analogous to those of genes regulated by Spx. These include putative antioxidant products (*ykuU*, *ykuV*, and *ydbD*) and short chain dehydrogenases (*ydaD*, *yhdF*, *yhxC,* and *yhxD*) that were proposed to function in the gluconate shunt that fuels oxidative pentose phosphate pathway for NADPH production, thus generating reducing power for reduction of oxidized thiols and for detoxifying enzymes. As *spx* is also transcriptionally upregulated by active σ^B, there exists the potential for activation of to regulons of specifying similar functions. It is notable that the regulons are distinct, possibly a reflection of differences in DNA target recognition by the two paralogs and/or in RNA holoenzyme specificity (Spx/σ^A and MgsR/σ^B).

A recent study explored the Spx regulon and its function in *Streptococcus sanguinis* through microarray hybridization analysis of *S. sanguinis* RNA (Chen et al. 2012). As with other members of *Firmicutes*, *S. sanguinis* bears two paralogs of *spx*, *spxA1*, and *spxA2*. Mutations of each affected oxidative stress tolerance and heat sensitivity, while *spxA1* mutant affected virulence. Microarray analysis showed reduced levels of transcripts specifying thioredoxin, superoxide dismutase, and thiol peroxidases. Interestingly, the genes whose products catalyze H_2O_2 production, *spxB*, and *nox* also required SpxA1 for optimal expression. *spxB* and *nox* encode pyruvate oxidase and H_2O_2-generating NADH oxidase, respectively. The *spxA1* mutant, although able to grow at a rate resembling that of the wild type parent, was unable to compete in a co-culture with *S. mutans*, presumably due to the inability to produce $H_2O_2.$

Our laboratory had recently investigated the composition of the Spx regulons in *B. anthracis* Sterne strain 7702. Using interspecies conjugation mediated by the *B. subtilis* ICEBs1 element (Auchtung et al. 2005), IPTG-inducible mutant alleles of *B. anthracis spxA1* and *spxA2* encoding protease resistant products (SpxA1^{DD} and SpxA2^{DD}) were introduced into *B. anthracis* (S. Barendt, M. M. Nakano, H. Lee, C. Birch, and P. Z, unpublished). Transcripts showing elevated concentrations following induction of mutant *spx* allele expression were identified by microarray hybridization technology. Some overlap was observed between the SpxA1 and

SpxA2 regulons, with *trxA*, *trxB*, and genes whose products function in methionine and cysteine biosynthesis induced by protease-resistant SpxA1 and SpxA2. Several group (*B. cereus/thuringiensis/anthracis*)-specific genes were among those induced by SpxA1 and A2. A gene not found in *B. subtilis* and implicated in redox homeostasis in *B. anthracis* and *S. aureus* is the *cdrA* gene encoding coenzyme A disulfide reductase (delCardayre et al. 1998; Luba et al. 1999). Spx-dependent control of *cdrA* was reconstructed in *B. subtilis* using a *cdrA-lacZ* fusion and expression of protease resistant forms of SpxA1 and A2 within a Δ*spx* mutant background. In studies of *B. anthracis* and *S. aureus*, CoADR has been found to play a role in maintaining the reduced state of CoA and thiols that are oxidatively modified by CoA. It has been proposed that CoA along with bacillithiol function together as a redox buffer, performing the function in a way that is analogous to glutathione in Gram-negative and eukaryotic systems (Luba et al. 1999; Parsonage et al. 2010). Both SpxA1 and SpxA2 activate transcription of genes that function in DNA repair, including *uvrC* and *exoA*.

The *spx* genes of *B. anthracis* are induced soon after germination in the infected macrophage (Bergman et al. 2007), a time when the outgrowing cell is subject to the oxidative assault within the phagocyte. Several of the genes activated by SpxA1 and A2 function in antioxidant activities, potentially link SpxA function with *B. anthracis* virulence. Interestingly, SpxA2 stimulates the transcription of the *racE* gene (S. Barendt, M. M. Nakano, H. Lee, C. Birch, and P. Z, unpublished), encoding the glutamate racemase required for the generation of D-glutamate, a precursor of the poly-γ-glutamyl protective capsule. The transcriptomic analysis of *B. anthracis* implicates the SpxA paralogs in virulence related functions, which was reinforced by studies of Spx in pathogenic Gram-positive cocci (Chen et al. 2012; Kajfasz et al. 2010, 2012).

Spx Function and Bacterial Virulence

The studies of *spx* mutants and Spx-activated transcriptomes have linked the protein to antioxidant activities that are typically associated with bacterial-mediated pathogenesis. The two *spx* paralogs of *S. mutans* had been shown to be required for pH and ROS tolerance. Using the *Galleria mellonella* (waxworm) model as host, Lemos and coworkers demonstrated that application of *S. mutans* to *G. mellonella* larvae resulted in host lethality (Kajfasz et al. 2010). Bactericidal phagocytosis in the waxworm is carried out by hemocytes, which mimic neutrophil function. The injection of *spxA1*, *spxA2*, and *spxA1 spxA2* mutants resulted in 50–25% larvae survival, indicating a role of the Spx paralogs in pathogenesis by counteracting the oxidative assault mediated by larval hemocytes. Further investigation showed that the *spx* paralogs were required for colonization of rat teeth (Kajfasz et al. 2010), in keeping with the known involvement of *S. mutans* in the generation of dental carries.

Studies of the opportunistic pathogen, *Enterococcus faecalis* showed that Spx plays a critical role in pathogen survival during macrophage infection, and in establishment of implant-induced peritoneal infection (Kajfasz et al. 2012). An *spx* null mutant showed a three-log reduction in survival within murine macrophages. Colonization of the murine peritoneal cavity and spleen by *E. faecalis* was evident after 48 h following implant application (10^3–10^4 cfus recovered). However, the *spx* null mutant showed a three-log drop in numbers after infection, both in peritoneum and spleen. Analysis of the transcriptome of the *spx* mutant uncovered genes encoding glutathione reductase, methionine sulfoxide reductase, NADH peroxidases, organic hydroperoxide reductase, superoxide dismutase, thiol peroxidases and thioredoxin reductase as part of the Spx regulon in *E. faecalis*. The reduced antioxidant expression in the *spx* mutant would potentially render the mutant defective in coping with the oxidative environment of the blood and within professional phagocytic cells. Hence, this was thought to be the cause of the low numbers of mutant bacteria recovered from the murine implant model.

Recently, the role of *spx* in infective endocarditis (IE) caused by *Streptococcus sanguinis* was investigated (Chen et al. 2012). The study was accompanied by transcriptomic analysis to uncover the regulon governed by Spx. As with several species within *Firmicutes*, *S. sanguinis* bears two paralogs of *spx*. SpxA2 was necessary for optimal growth as well as resistance to low pH, high temperature, and oxidative stress. The *spxA1* null mutation conferred sensitivity to oxidative stress and blocked the production of H_2O_2. The latter effect was found to be due to reduced expression of *spxB* (pyruvate oxidase) and *nox* (NADH oxidase), the products of which function in peroxide production. The *spxA1*, but not the *spxA2*, null mutation caused a defect in IE generation. The authors of the study speculate that the defect in IE conferred by the *spxA1* mutation is due in part to the reduced peroxide production and antioxidant gene expression. Interestingly, the *spxA* mutant of *E. faecalis* also showed reduced *nox* gene expression, based on microarray transcriptomic analysis.

The results reported thus far suggest that Spx orthologs are intimately associated with the control of virulence determinants within Gram-positive organisms known to cause infectious disease. There is a considerable body of evidence from mutational and transcriptomic analysis that Spx governs the expression of genes that would be expected to play critical roles in pathogen survival in the host. Data from studies of *S. sanguinis* and *E. faecalis* suggest that genes (*spxB* and *nox*) whose products contribute directly to pathogenesis might also operate within the realm of Spx control.

Questions and Future Directions

Much of what is known about the role Spx plays in the stress response comes from the genome-wide analyses of Spx-dependent transcriptional control and from the chromatin immunoprecipitation experiments, which uncovered genes directly contacted by Spx/RNA polymerase. Collectively, the data has provided a glimpse of the role played by Spx in the stress response. Spx orthologs have been assigned functions within the cells' oxidative stress and cell envelope stress responses. However, in one species, *S. pneumoniae*, the paralogous *spx* genes, *spxA1* and *spxA2*, are essential. The reasons for why the double *spxA1 spxA2* mutant is non-viable have yet to be uncovered (Turlan et al. 2009). The inability to create the double mutant has hindered investigation of Spx in this organism, as the two paralogs appear to have overlapping realms of control, and creating a null in *spxA1* had little effect on oxidant resistance, presumably because of the compensatory role played by the SpxA2 paralog (Turlan et al. 2009).

The activation and accumulation of Spx resulting from redox stress causes temporary repression of genes that function in developmental processes, such as sporulation and competence (K state in *B. subtilis*, X state in *S. pneumoniae*). The prestress state is soon restored when redox stress is alleviated, which is accompanied by lower Spx-activated transcription, reduction in Spx concentration (Nakano et al. 2003a), and conversion of Spx to the inactive, reduced state (M. Nakano, and P. Z. unpublished). In light of the observed global repressive effects exerted by high Spx concentration, mechanisms responsible for reducing Spx concentration and activity to prestress levels would seem to be crucial, yet there is currently little information about how they operate.

It is known that Spx concentration is controlled at the level of proteolysis that is catalyzed by ClpXP and in some organisms, enhanced by YjbH, but how is this activity controlled to allow accumulation of Spx in times of stress, and to eliminate Spx when stress subsides? It is not known what serves as the sensory component of this system of regulated proteolysis. One can imagine that Spx itself is the sensor, not only for its activity but also its proteolytic turnover. While no difference in sensitivity to ClpXP between reduced and oxidized Spx was observed (Zhang and Zuber 2007), oxidized Spx, through its interaction with RNA polymerase, might be protected from YjbH-mediated proteolysis. Data from two studies (Chan et al. 2012; Engman et al. 2012) suggest that YjbH does not serve a sensory role, although it is possible that it might be the target of another factor that is responsive to the cellular redox conditions.

Figure 5 shows a speculative model of post-translational Spx control in response to oxidative stress. Spx in oxidized form is shown to interact with RNA polymerase. One component of proteolytic Spx control that could serve a sensory role is the ClpX protein. Previous work from our lab showed that ClpXP could be inactivated by treatment with peroxide or diamide. Sensitivity to oxidants might be attributed to a redox sensitive, N-terminal Zn-binding domain, in which a zinc atom is coordinated by four Cys residues (Fig. 5). Oxidative inactivation of ClpX would result in increase in Spx concentration, and increase in the number of Spx/RNA polymerase complexes. Figure 6 shows a model of how Spx concentration is reduced upon recovery from stress. The systems activated by Spx serve to convert Spx to its inactive reduced state, releasing Spx from RNA polymerase and rendering it susceptible to YjbH-dependent proteolysis by ClpXP. Recent studies have provided evidence that the *clpX* gene is positively controlled by activated Spx (Rochat et al. 2012), suggesting that redox stress generates heightened concentrations of ClpX protein (Fig. 6). This might serve as a possible measure within the

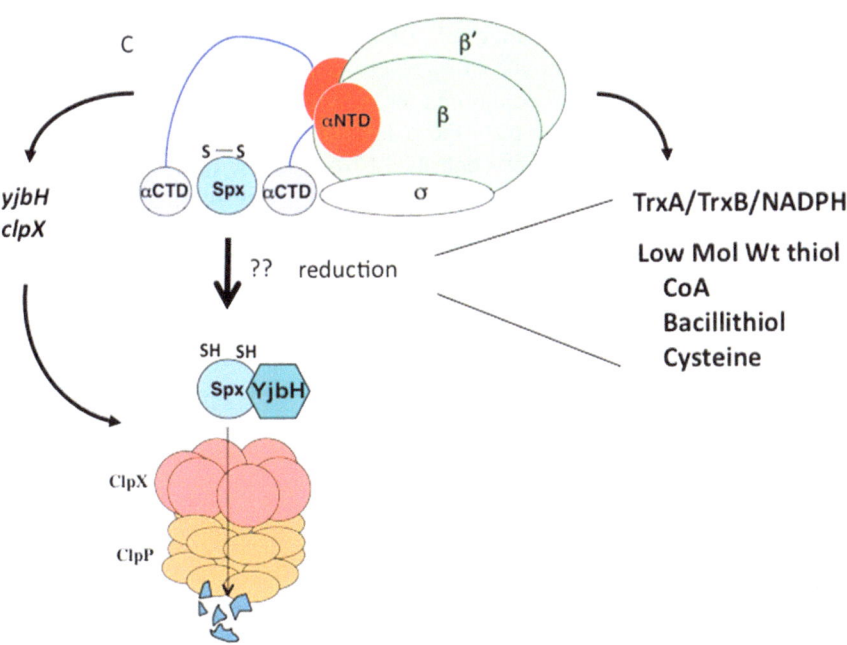

Fig. 6 Spx control and recovery from stress. Spx activates the transcription of genes whose products function in thiol redox homeostasis, and genes that function in Spx degradation. Reduced Spx loses affinity for RNA polymerase and forms a complex with YjbH, now at elevated concentrations. Spx is degraded by ClpXP, which is formed by newly synthesized ClpX protein

cell to ensure that enough ClpXP is in place to rapidly dispose of Spx when redox stress subsides. Questions arise as to whether the N-terminal Zn-binding domain of ClpX undergoes a change in its redox state upon oxidative stress, how this might affect activity, and if such a change in redox state, if it occurs, is reversible.

Evidence obtained thus far has supported the hypothesis that Spx is a sensor of cellular redox state. While it is known that oxidation of Spx results in productive interaction of the protein with RNA polymerase (Fig. 5), the detailed structural description of how oxidized Spx contacts RNA polymerase and how this leads to transcriptional activation has not been achieved. Studies using thiol-reactive agents and immunoblot analysis have shown that the redox state of Spx changes rapidly from oxidized to reduced when redox homeostasis is restored (M. M. Nakano and P. Z. unpublished). The restoration of Spx to the reduced inactive state depends on Spx/RNA polymerase interaction, since the *rpoAY263C* mutation conferring a defect in Spx-RNA polymerase contact also results in prolonged periods of high Spx concentration following diamide-induced stress (Nakano et al. 2003a). A number of factors could facilitate or catalyze this transition, several of which are probably under Spx control (Fig. 6). These might include NADPH/thioredoxin reductase/thioredoxin system for maintaining thiols in their reduced state. This system is also under σ^B control, which directs transcription from one of the promoters of the *trxA* gene (Scharf et al. 1998). This illustrates another potential avenue of Spx integration into the stress response network. Reduction of oxidized Spx could also be accomplished through the activity of the enzymes that catalyze bacillithiol biosynthesis and control the redox state of the low molecular weight biothiol. In *B. anthracis*, the Spx-controlled CoA disulfide reductases might also function directly or indirectly in determining the redox state of SpxA1 or A2.

A thorough understanding of how Spx ortholog function is integrated in global transcriptional control will require investigations into the relationship of Spx with the various forms of RNA polymerase holoenzyme and with other transcriptional regulators. There have been few studies of Spx interaction with alternative forms of RNA polymerase. Previous studies suggested that certain holoenzyme forms might be inhibited by Spx, but it is not known if Spx can interact productively with forms of holoenzyme other than that which bears the major, essential sigma subunit, which in *B. subtilis* is σ^A. Active σ^X, σ^W, and σ^M all serve to induce *spx* transcription, but might the Spx protein that is generated now engage these holoenzyme forms to active a unique regulon of genes? Examination of promoter architecture has not, so far, uncovered likely candidates as being utilized by a complex of Spx interacting with a holoenzyme bearing one of the cell envelope stress σ subunits, however. The *mgsR* gene, a member of the σ^B regulon, is positively autoregulated by its product (Reder et al. 2012b), and future studies will likely examine the interaction between the Spx paralog and the σ^B RNA polymerase holoenzyme. What structural features of RNA polymerase, Spx, and promoter anatomy might converge to facilitate the formation of productive complexes between Spx and alternative RNA polymerase forms? There is no evidence to date that Spx or its orthologs can interact with σ protein, although evidence indicates that Spx has higher affinity for holo- rather than core enzyme (Lin and Zuber 2012). Promoter elements that incorporate core sequences recognized by the σ subunit along with the upstream sequence required for Spx control might be all that is required to generate an active Spx/RNA polymerase transcription initiation complex.

What is the fate of Spx after transcription initiation? The interaction of Spx with RNA polymerase holoenzyme might also persist past the initiation stage of the transcription process. There is no evidence reported that has indicated if/when Spx is released from the transcription complex. Potentially, Spx could exert control over elongation and/or termination.

Could Spx interaction with RNA polymerase provide an additional platform for communication with transcriptional regulators? Recent studies of *B. subtilis* involving an investigation into the response to hypochlorite uncovered a regulator, HypR (Palm et al. 2012), a MarR/Duf24 family member that controls genes whose products function in alleviating disulfide stress and in detoxification reactions. Many of these genes are also controlled by SpxA, which raises the question of whether HypR can activate transcription by engaging Spx/RNA polymerase to generate a transcription initiation complex. A number of genes encoding transcriptional regulators are induced through an Spx-dependent mechanism. These include genes (e.g., *ytzE*) that are also induced by the stringent response, by phosphate starvation (PhoR-independent), and by disulfide stress (Allenby et al. 2005; Eymann et al. 2002; Petersohn et al. 2001). For most of these genes that encode members of known transcription factor families, there is little to no information regarding the genes they control or the conditions (or specific ligands) that lead to their activation. One scenario of stress-responsive transcriptional control might involve the product of an Spx-activated gene serving to recruit Spx/RNA polymerase to establish a transcription initiation complex. In this case, Spx would serve as an RNA polymerase subunit, rather than a sequence-specific DNA-binding component of the transcription complex. Thus, there could exist "hidden" Spx regulons yet to be uncovered that are co-regulated by Spx-interacting transcription factors.

The picture that emerges of Spx control is that a variety of stress conditions induce the expression of *spx*, but changes in the intracellular redox status affect its activity and post-translationally regulate its concentration. The conditions that induce *spx* elicit responses that involve the activation of alternative RNA polymerase holoenzymes, TCSTSs, and mechanisms of derepression that stimulate *spx* expression. Because Spx becomes a subunit of RNA polymerase, it can potentially exert control, positive and negative, over a broad spectrum of cellular processes. These include complex programs of cellular differentiation and cell specialization, systems of oxidative stress mitigation, and transcriptional control of metabolic processes. Further understanding its integration into the stress response network will require greater knowledge of its relationship with RNA polymerase and other transcription factors, and a deeper understanding of the genes that it controls in the various Gram-positive species.

From the phenotype of *spx* mutants, the influence of Spx over the bacterial transcriptome and its function in stress response control, Spx has gained some importance as a virulence determinant. Evidence accumulated thus far indicates that it is solely prokaryotic in origin, and primarily within Gram-positive species. Hence, opportunities very likely exist for the development of effective antibacterial agents that target Spx activity.

Acknowledgments My thanks to Charles P. Moran, Jr., Michiko M. Nakano, Catherine J. Kemp, and Skye Barendt for reading the manuscript and providing helpful comments and suggestions.

References

Akbar S, Gaidenko TA, Kang CM, O'Reilly M, Devine KM, Price CW (2001) New family of regulators in the environmental signaling pathway which activates the general stress transcription factor sigma(B) of *Bacillus subtilis*. J Bacteriol 183:1329–1338

Akbar S, Kang CM, Gaidenko TA, Price CW (1997) Modulator protein RsbR regulates environmental signalling in the general stress pathway of *Bacillus subtilis*. Mol Microbiol 24:567–578

Allenby NE, O'Connor N, Pragai Z, Ward AC, Wipat A, Harwood CR (2005) Genome-wide transcriptional analysis of the phosphate starvation stimulon of *Bacillus subtilis*. J Bacteriol 187:8063–8080

Anjem A, Imlay JA (2012) Mononuclear iron enzymes are primary targets of hydrogen peroxide stress. J Biol Chem 287:15544–15556

Antelmann H, Scharf C, Hecker M (2000) Phosphate starvation-inducible proteins of *Bacillus subtilis*: proteomics and transcriptional analysis. J Bacteriol 182:4478–4490

Auchtung JM, Lee CA, Monson RE, Lehman AP, Grossman AD (2005) Regulation of a *Bacillus subtilis* mobile genetic element by intercellular signaling and the global DNA damage response. Proc Natl Acad Sci USA 102:12554–12559

Baldus JM, Buckner CM, Moran CPJ (1995) Evidence that the transcriptional activator Spo0A interacts with two sigma factors in *Bacillus subtilis*. Mol Microbiol 17:281–290

Banse AV, Hobbs EC, Losick R (2011) Phosphorylation of Spo0A by the histidine kinase KinD requires the lipoprotein *med* in *Bacillus subtilis*. J Bacteriol 193:3949–3955

Bergman NH, Anderson EC, Swenson EE, Janes BK, Fisher N, Niemeyer MM, Miyoshi AD, Hanna PC (2007) Transcriptional profiling of *Bacillus anthracis* during infection of host macrophages. Infect Immun 75:3434–3444

Bergman NH, Anderson EC, Swenson EE, Niemeyer MM, Miyoshi AD, Hanna PC (2006) Transcriptional profiling of the *Bacillus anthracis* life cycle in vitro and an implied model for regulation of spore formation. J Bacteriol 188:6092–6100

Berka RM, Hahn J, Albano M, Draskovic I, Persuh M, Cui X, Sloma A, Widner W, Dubnau D (2002) Microarray analysis of the *Bacillus subtilis* K-state: genome-wide expression changes dependent on ComK. Mol Microbiol 43:1331–1345

Borezee E, Msadek T, Durant L, Berche P (2000) Identification in *Listeria monocytogenes* of MecA, a homologue of the *Bacillus subtilis* competence regulatory protein. J Bacteriol 182:5931–5934

Brody MS, Vijay K, Price CW (2001) Catalytic function of an alpha/beta hydrolase is required for energy stress activation of the sigma(B) transcription factor in *Bacillus subtilis*. J Bacteriol 183:6422–6428

Buchko GW, Hewitt SN, Napuli AJ, Van Voorhis WC, Myler PJ (2011) Solution structure of an arsenate reductase-related protein, YffB, from *Brucella melitensis*, the etiological agent responsible for brucellosis. Acta Crystallogr Sect F Struct Biol Cryst Commun 67:1129–1136

P. Zuber, *Function and Control of the Spx-Family of Proteins Within the Bacterial Stress Response*, SpringerBriefs in Microbiology, DOI: 10.1007/978-1-4614-6925-4,
© The Author(s) 2013

Burbulys D, Trach KA, Hoch JA (1991) Initiation of sporulation in *B. subtilis* is controlled by a multicomponent phosphorelay. Cell 64:545–552

Cao M, Helmann JD (2004) The *Bacillus subtilis* extracytoplasmic-function sigmaX factor regulates modification of the cell envelope and resistance to cationic antimicrobial peptides. J Bacteriol 186:1136–1146

Cao M, Kobel PA, Morshedi MM, Wu MF, Paddon C, Helmann JD (2002) Defining the *Bacillus subtilis* sigma(W) regulon: a comparative analysis of promoter consensus search, run-off transcription/macroarray analysis (ROMA), and transcriptional profiling approaches. J Mol Biol 316:443–457

Cao M, Wang T, Ye R, Helmann JD (2002) Antibiotics that inhibit cell wall biosynthesis induce expression of the *Bacillus subtilis* sigma(W) and sigma(M) regulons. Mol Microbiol 45:1267–1276

Chan CM, Garg S, Lin AA, Zuber P (2012) Geobacillus thermodenitrificans YjbH recognizes the C-terminal end of *Bacillus subtilis* Spx to accelerate Spx proteolysis by ClpXP. Microbiology 158:1268–1278

Chaturongakul S, Boor KJ (2004) RsbT and RsbV contribute to sigmaB-dependent survival under environmental, energy, and intracellular stress conditions in *Listeria monocytogenes*. Appl Environ Microbiol 70:5349–5356

Chen L, Ge X, Wang X, Patel JR, Xu P (2012) SpxA1 involved in hydrogen peroxide production, stress tolerance and endocarditis virulence in *Streptococcus sanguinis*. PLoS ONE 7:e40034

Chi BK, Gronau K, Maeder U, Hessling B, Becher D, Antelmann H (2011) S-bacillithiolation protects against hypochlorite stress in *Bacillus subtilis* as revealed by transcriptomics and redox proteomics. Mol Cell Proteomics 10:M111.009506

Chi BK, Kobayashi K, Albrecht D, Hecker M, Antelmann H (2010) The paralogous MarR/DUF24-family repressors YodB and CatR control expression of the catechol dioxygenase CatE in *Bacillus subtilis*. J Bacteriol 192:4571–4581

Christensen QH, Martin N, Mansilla MC, de Mendoza D, Cronan JE (2011) A novel amidotransferase required for lipoic acid cofactor assembly in *Bacillus subtilis*. Mol Microbiol 80:350–363

Cue D, Lei MG, Lee CY (2012) Genetic regulation of the intercellular adhesion locus in staphylococci. Front Cell Infect Microbiol 2:38

D'Souza C, Nakano MM, Zuber P (1994) Identification of *comS*, a gene of the *srfA* operon that regulates the establishment of genetic competence in *Bacillus subtilis*. Proc Natl Acad Sci USA 91:9397–9401

de Been M, Tempelaars MH, van Schaik W, Moezelaar R, Siezen RJ, Abee T (2010) A novel hybrid kinase is essential for regulating the sigma(B)-mediated stress response of *Bacillus cereus*. Environ Microbiol 12:730–745

de Hoon MJ, Eichenberger P, Vitkup D (2010) Hierarchical evolution of the bacterial sporulation network. Curr Biol 20:R735–745

delCardayre SB, Stock KP, Newton GL, Fahey RC, Davies JE (1998) Coenzyme A disulfide reductase, the primary low molecular weight disulfide reductase from *Staphylococcus aureus*. Purification and characterization of the native enzyme. J Biol Chem 273:5744–5751

Dintner S, Staroń A, Berchtold E, Petri T, Mascher T, Gebhard S (2011) Coevolution of ABC transporters and two-component regulatory systems as resistance modules against antimicrobial peptides in firmicutes bacteria. J Bacteriol 193:3851–3862

Duncan L, Alper S, Losick R (1994) Establishment of cell type specific gene transcription during sporulation in *Bacillus subtilis*. Curr Opin Genet Dev 4:630–636

Duwat P, Ehrlich SD, Gruss A (1999) Effects of metabolic flux on stress response pathways in *Lactococcus lactis*. Mol Microbiol 31:845–858

Eiamphungporn W, Helmann JD (2008) The *Bacillus subtilis* sigma(M) regulon and its contribution to cell envelope stress responses. Mol Microbiol 67:830–848

Eldholm V, Gutt B, Johnsborg O, Bruckner R, Maurer P, Hakenbeck R, Mascher T, Havarstein LS (2010) The pneumococcal cell envelope stress-sensing system LiaFSR is activated by murein hydrolases and lipid II-interacting antibiotics. J Bacteriol 192:1761–1773

Engman J, Rogstam A, Frees D, Ingmer H, von Wachenfeldt C (2012) The YjbH adaptor protein enhances proteolysis of the transcriptional regulator Spx in *Staphylococcus aureus*. J Bacteriol 194:1186–1194

Errington J (1993) *Bacillus* sporulation: regulation of gene expression and control of morphogenesis. Microbiol Rev 57:1–33

Even S, Burguiere P, Auger S, Soutourina O, Danchin A, Martin-Verstraete I (2006) Global control of cysteine metabolism by CymR in *Bacillus subtilis*. J Bacteriol 188:2184–2197

Eymann C, Homuth G, Scharf C, Hecker M (2002) *Bacillus subtilis* functional genomics: global characterization of the stringent response by proteome and transcriptome analysis. J Bacteriol 184:2500–2520

Faulkner MJ, Ma Z, Fuangthong M, Helmann JD (2012) Derepression of the *Bacillus subtilis* PerR peroxide stress response leads to iron deficiency. J Bacteriol 194:1226–1235

Ferreira A, Gray M, Wiedmann M, Boor KJ (2004) Comparative genomic analysis of the *sigB* operon in *Listeria monocytogenes* and in other Gram-positive bacteria. Curr Microbiol 48:39–46

Formstone A, Carballido-Lopez R, Noirot P, Errington J, Scheffers DJ (2008) Localization and interactions of teichoic acid synthetic enzymes in *Bacillus subtilis*. J Bacteriol 190:1812–1821

Fredrickson JK, Li SM, Gaidamakova EK, Matrosova VY, Zhai M, Sulloway HM, Scholten JC, Brown MG, Balkwill DL, Daly MJ (2008) Protein oxidation: key to bacterial desiccation resistance? ISME J 2:393–403

Fuangthong M, Atichartpongkul S, Mongkolsuk S, Helmann JD (2001) OhrR is a repressor of *ohrA*, a key organic hydroperoxide resistance determinant in *Bacillus subtilis*. J Bacteriol 183:4134–4141

Fuangthong M, Herbig AF, Bsat N, Helmann JD (2002) Regulation of the *Bacillus subtilis fur* and *perR* genes by PerR: not all members of the PerR regulon are peroxide inducible. J Bacteriol 184:3276–3286

Fujita M, Gonzalez-Pastor JE, Losick R (2005) High-and low-threshold genes in the Spo0A regulon of *Bacillus subtilis*. J Bacteriol 187:1357–1368

Fujita M, Losick R (2005) Evidence that entry into sporulation in *Bacillus subtilis* is governed by a gradual increase in the level and activity of the master regulator Spo0A. Genes Dev 19:2236–2244

Gaballa A, Newton GL, Antelmann H, Parsonage D, Upton H, Rawat M, Claiborne A, Fahey RC, Helmann JD (2010) Biosynthesis and functions of bacillithiol, a major low-molecular-weight thiol in bacilli. Proc Natl Acad Sci USA 107:6482–6486

Gaidenko TA, Kim TJ, Weigel AL, Brody MS, Price CW (2006) The blue-light receptor YtvA acts in the environmental stress signaling pathway of *Bacillus subtilis*. J Bacteriol 188:6387–6395

Garg SK, Kommineni S, Henslee L, Zhang Y, Zuber P (2009) The YjbH protein of *Bacillus subtilis* enhances ClpXP-catalyzed proteolysis of Spx. J Bacteriol 191:1268–1277

Gebhard S (2012) ABC transporters of antimicrobial peptides in firmicutes bacteria—phylogeny, function and regulation. Mol Microbiol 86:1295–1317

Gonzalez-Pastor JE, Hobbs EC, Losick R (2003) Cannibalism by sporulating bacteria. Science 301:510–513

Grossman AD (1995) Genetic networks controlling the initiation of sporulation and the development of genetic competence in *Bacillus subtilis*. Ann Rev Genet 29:477–508

Hamoen LW, Eshuis H, Jongbloed J, Venema G, van Sinderen D (1995) A small gene, designated *comS*, located within the coding region of the fourth amino acid-activation domain of *srfA*, is required for competence development in *Bacillus subtilis*. Mol Microbiol 15:55–63

Hamoen LW, Venema G, Kuipers OP (2003) Controlling competence in *Bacillus subtilis*: shared use of regulators. Microbiology 149:9–17

Harwood CR, Coxon RD, Hancock IC (1990) The *Bacillus* cell envelope and secretion. In: Harwood CR, Cutting SM (eds) Molecular biological methods for *Bacillus*. John Wiley & Sons, Chichester, pp 327–390

Helmann JD (2011) Bacillithiol, a new player in bacterial redox homeostasis. Antioxid Redox Signal 15:123–133

Helmann JD, Wu MF, Kobel PA, Gamo FJ, Wilson M, Morshedi MM, Navre M, Paddon C (2001) Global transcriptional response of *Bacillus subtilis* to heat shock. J Bacteriol 183:7318–7328

Hengge R (2011) The general stress response in gram-negative bacteria. In: Storz G, Hengge R (eds) Bacterial stress responses. ASM, Washington, DC, pp 251–290

Hengge-Aronis R (1993) Survival of hunger and stress: the role of *rpoS* in stationary phase gene regulation in *Escherichia coli*. Cell 72:165–168

Hersch GL, Baker TA, Sauer RT (2004) SspB delivery of substrates for ClpXP proteolysis probed by the design of improved degradation tags. Proc Natl Acad Sci USA 101:12136–12141

Hoch JA (1993) Regulation of the phosphorelay and the initiation of sporulation in *Bacillus subtilis*. Annu Rev Microbiol 47:441–465

Hoch JA (2000) Two-component and phosphorelay signal transduction. Curr Opin Microbiol 3:165–170

Hochgrafe F, Wolf C, Fuchs S, Liebeke M, Lalk M, Engelmann S, Hecker M (2008) Nitric oxide stress induces different responses but mediates comparable protein thiol protection in *Bacillus subtilis* and *Staphylococcus aureus*. J Bacteriol 190:4997–5008

Hofmeister AEM, Londono-Vallejo A, Harry E, Stragier P, Losick R (1995) Extracellular signal protein triggering the proteolytic activation of a developmental transcription factor in *B. subtilis*. Cell 83:219–226

Horsburgh MJ, Clements MO, Crossley H, Ingham E, Foster SJ (2001) PerR controls oxidative stress resistance and iron storage proteins and is required for virulence in *Staphylococcus aureus*. Infect Immun 69:3744–3754

Huang X, Gaballa A, Cao M, Helmann JD (1999) Identification of target promoters for the *Bacillus subtilis* extracytoplasmic function sigma factor, sigma W. Mol Microbiol 31:361–371

Hyyrylainen HL, Pietiainen M, Lunden T, Ekman A, Gardemeister M, Murtomaki-Repo S, Antelmann H, Hecker M, Valmu L, Sarvas M, Kontinen VP (2007) The density of negative charge in the cell wall influences two-component signal transduction in *Bacillus subtilis*. Microbiology 153:2126–2136

Imlay JA (2003) Pathways of oxidative damage. Annu Rev Microbiol 57:395–418

Imlay JA (2006) Iron-sulphur clusters and the problem with oxygen. Mol Microbiol 59:1073–1082

Imlay JA, Linn S (1988) DNA damage and oxygen radical toxicity. Science 240:1302–1309

Jervis AJ, Thackray PD, Houston CW, Horsburgh MJ, Moir A (2007) SigM-responsive genes of *Bacillus subtilis* and their promoters. J Bacteriol 189:4534–4538

Jiang M, Shao W, Perego M, Hoch JA (2000) Multiple histidine kinases regulate entry into stationary phase and sporulation in *Bacillus subtilis*. Mol Microbiol 38:535–542

Jordan S, Hutchings MI, Mascher T (2008) Cell envelope stress response in Gram-positive bacteria. FEMS Microbiol Rev 32:107–146

Kajfasz JK, Martinez AR, Rivera-Ramos I, Abranches J, Koo H, Quivey RG Jr, Lemos JA (2009) Role of Clp proteins in expression of virulence properties of *Streptococcus mutans*. J Bacteriol 191:2060–2068

Kajfasz JK, Mendoza JE, Gaca AO, Miller JH, Koselny KA, Giambiagi-Demarval M, Wellington M, Abranches J, Lemos JA (2012) The Spx regulator modulates stress responses and virulence in *Enterococcus faecalis*. Infect Immun 80:2265–2275

Kajfasz JK, Rivera-Ramos I, Abranches J, Martinez AR, Rosalen PL, Derr AM, Quivey RG, Lemos JA (2010) Two Spx proteins modulate stress tolerance, survival, and virulence in *Streptococcus mutans*. J Bacteriol

Kang CM, Vijay K, Price CW (1998) Serine kinase activity of a *Bacillus subtilis* switch protein is required to transduce environmental stress signals but not to activate its target PP2C phosphatase. Mol Microbiol 30:189–196

Karow ML, Glaser P, Piggot PJ (1995) Idenitification of a gene, *spoIIR*, that links the activation of σ^E to the transcriptional activit of σ^F during sporulation in *Bacillus subtilis*. Proc Natl Acad Sci 92:2012–2016

Keren IY, Wu JI, Mulcahy LR, Lewis K (2013) Killing by bactericidal antibiotics does not depend on reactive oxygen species. Science 339:1213–1216

Kim MS, Hahn MY, Cho Y, Cho SN, Roe JH (2009) Positive and negative feedback regulatory loops of thiol-oxidative stress response mediated by an unstable isoform of sigmaR in actinomycetes. Mol Microbiol 73:815–825

Kirstein J, Moliere N, Dougan DA, Turgay K (2009) Adapting the machine: adaptor proteins for Hsp100/Clp and AAA+ proteases. Nat Rev Microbiol 7:589–599

Kohanski MA, Dwyer DJ, Collins JJ (2010) How antibiotics kill bacteria: from targets to networks. Nat Rev Microbiol 8:423–435

Kohanski MA, Dwyer DJ, Hayete B, Lawrence CA, Collins JJ (2007) A common mechanism of cellular death induced by bactericidal antibiotics. Cell 130:797–810

Kommineni S, Garg SK, Chan CM, Zuber P (2011) YjbH-enhanced proteolysis of Spx by ClpXP in *Bacillus subtilis* is inhibited by the small protein YirB (YuzO). J Bacteriol 193:2133–2140

Kristian SA, Datta V, Weidenmaier C, Kansal R, Fedtke I, Peschel A, Gallo RL, Nizet V (2005) D-Alanylation of teichoic acids promotes group a *streptococcus* antimicrobial peptide resistance, neutrophil survival, and epithelial cell invasion. J Bacteriol 187:6719–6725

Larsson JT, Rogstam A, von Wachenfeldt C (2007) YjbH is a novel negative effector of the disulphide stress regulator, Spx, in *Bacillus subtilis*. Mol Microbiol 66:669–684

Leelakriangsak M, Huyen NT, Towe S, van Duy N, Becher D, Hecker M, Antelmann H, Zuber P (2008) Regulation of quinone detoxification by the thiol stress sensing DUF24/MarR-like repressor, YodB in *Bacillus subtilis*. Mol Microbiol 67:1108–1124

Leelakriangsak M, Kobayashi K, Zuber P (2007) Dual negative control of *spx* transcription initiation from the P3 promoter by repressors PerR and YodB in *Bacillus subtilis*. J Bacteriol 189:1736–1744

Leelakriangsak M, Zuber P (2007) Transcription from the P3 promoter of the *Bacillus subtilis* *spx* gene is induced in response to disulfide stress. J Bacteriol 189:1727–1735

Leichert LI, Scharf C, Hecker M (2003) Global characterization of disulfide stress in *Bacillus subtilis*. J Bacteriol 185:1967–1975

Lin AA, Zuber P (2012) Evidence that a single monomer of Spx can productively interact with RNA polymerase in *Bacillus subtilis*. J Bacteriol 194:1697–1707

Liu H, Bergman NH, Thomason B, Shallom S, Hazen A, Crossno J, Rasko DA, Ravel J, Read TD, Peterson SN, Yates J 3rd, Hanna PC (2004) Formation and composition of the *Bacillus anthracis* endospore. J Bacteriol 186:164–178

Liu J, Cosby WM, Zuber P (1999) Role of Lon and ClpX in the post-translational regulation of a sigma subunit of RNA polymerase required for cellular differentiation of *Bacillus subtilis*. Mol Microbiol 33:415–428

Liu J, Zuber P (2000) The ClpX protein of *Bacillus subtilis* indirectly influences RNA polymerase holoenzyme composition and directly stimulates sigmaH-dependent transcription. Mol Microbiol 37:885–897

Liu Y, Imlay JA (2013) Cell death from antibiotics without the involvement of reactive oxygen species. Science 339:1210–1213

Luba J, Charrier V, Claiborne A (1999) Coenzyme A-disulfide reductase from *Staphylococcus aureus*: evidence for asymmetric behavior on interaction with pyridine nucleotides. Biochemistry 38:2725–2737

Luo Y, Asai K, Sadaie Y, Helmann JD (2010) Transcriptomic and phenotypic characterization of *Bacillus subtilis* strain without extracytoplasmic function sigma factors. J Bacteriol 192:5736–5745

Maamar H, Dubnau D (2005) Bistability in the *Bacillus subtilis* K-state (competence) system requires a positive feedback loop. Mol Microbiol 56:615–624

Maamar H, Raj A, Dubnau D (2007) Noise in gene expression determines cell fate in *Bacillus subtilis*. Science 317:526–529

Magnuson R, Solomon J, Grossman AD (1994) Biochemical and genetic characterization of a competence pheromone from *Bacillus subtilis*. Cell 77:207–216

Marles-Wright J, Grant T, Delumeau O, van Duinen G, Firbank SJ, Lewis PJ, Murray JW, Newman JA, Quin MB, Race PR, Rohou A, Tichelaar W, van Heel M, Lewis RJ (2008) Molecular architecture of the "stressosome," a signal integration and transduction hub. Science 322:92–96

Martin P, DeMel S, Shi J, Gladysheva T, Gatti DL, Rosen BP, Edwards BF (2001) Insights into the structure, solvation, and mechanism of ArsC arsenate reductase, a novel arsenic detoxification enzyme. Structure 9:1071–1081

Mascher T (2006) Intramembrane-sensing histidine kinases: a new family of cell envelope stress sensors in firmicutes bacteria. FEMS Microbiol Lett 264:133–144

Mascher T, Hachmann AB, Helmann JD (2007) Regulatory overlap and functional redundancy among *Bacillus subtilis* extracytoplasmic function sigma factors. J Bacteriol 189:6919–6927

Mascher T, Margulis NG, Wang T, Ye RW, Helmann JD (2003) Cell wall stress responses in *Bacillus subtilis*: the regulatory network of the bacitracin stimulon. Mol Microbiol 50:1591–1604

Messner KR, Imlay JA (1999) The identification of primary sites of superoxide and hydrogen peroxide formation in the aerobic respiratory chain and sulfite reductase complex of *Escherichia coli*. J Biol Chem 274:10119–10128

Michel A, Agerer F, Hauck CR, Herrmann M, Ullrich J, Hacker J, Ohlsen K (2006) Global regulatory impact of ClpP protease of *Staphylococcus aureus* on regulons involved in virulence, oxidative stress response, autolysis, and DNA repair. J Bacteriol 188:5783–5796

Minnig K, Barblan JL, Kehl S, Moller SB, Mauel C (2003) In *Bacillus subtilis* W23, the duet sigmaX sigmaM, two sigma factors of the extracytoplasmic function subfamily, are required for septum and wall synthesis under batch culture conditions. Mol Microbiol 49:1435–1447

Minnig K, Lazarevic V, Soldo B, Mauel C (2005) Analysis of teichoic acid biosynthesis regulation reveals that the extracytoplasmic function sigma factor sigmaM is induced by phosphate depletion in *Bacillus subtilis* W23. Microbiology 151:3041–3049

Mishra S, Imlay J (2012) Why do bacteria use so many enzymes to scavenge hydrogen peroxide? Arch Biochem Biophys. http://dx.doi.org/10.1016/j.abb.2012.04.014.

Mols M, Abee T (2011) Primary and secondary oxidative stress in *Bacillus*. Environ Microbiol 13:1387–1394

Murray JW, Delumeau O, Lewis RJ (2005) Structure of a nonheme globin in environmental stress signaling. Proc Natl Acad Sci USA 102:17320–17325

Nagorska K, Hinc K, Strauch MA, Obuchowski M (2008) Influence of the sigmaB stress factor and *yxaB*, the gene for a putative exopolysaccharide synthase under sigmaB control, on biofilm formation. J Bacteriol 190:3546–3556

Nakano MM, Hajarizadeh F, Zhu Y, Zuber P (2001) Loss-of-function mutations in *yjbD* result in ClpX- and ClpP-independent competence development of *Bacillus subtilis*. Mol Microbiol 42:383–394

Nakano MM, Lin A, Zuber CS, Newberry KJ, Brennan RG, Zuber P (2010) Promoter recognition by a complex of Spx and the C-terminal domain of the RNA polymerase Œ± subunit. PLoS ONE 5:e8664

Nakano MM, Zhu Y, Liu J, Reyes DY, Yoshikawa H, Zuber P (2000) Mutations conferring amino acid residue substitutions in the carboxy-terminal domain of RNA polymerase α can suppress *clpX* and *clpP* with respect to developmentally regulated transcription in *Bacillus subtilis*. Mol Microbiol 37:869–884

Nakano MM, Zuber P (1991) The primary role of *comA* in the establishment of the competent state in *Bacillus subtilis* is to activate the expression of *srfA*. J Bacteriol 173:7269–7274

Nakano S, Erwin KN, Ralle M, Zuber P (2005) Redox-sensitive transcriptional control by a thiol/disulphide switch in the global regulator. Spx Mol Microbiol 55:498–510

Nakano S, Küster-Schöck E, Grossman AD, Zuber P (2003) Spx-dependent global transcriptional control is induced by thiol-specific oxidative stress in *Bacillus subtilis*. Proc Natl Acad Sci USA 100:13603–13608

Nakano S, Nakano MM, Zhang Y, Leelakriangsak M, Zuber P (2003) A regulatory protein that interferes with activator-stimulated transcription in bacteria. Proc Natl Acad Sci USA 100:4233–4238

Nakano S, Zheng G, Nakano MM, Zuber P (2002) Multiple pathways of Spx (YjbD) proteolysis in *Bacillus subtilis*. J Bacteriol 184:3664–3670

Newberry KJ, Nakano S, Zuber P, Brennan RG (2005) Crystal structure of the *Bacillus subtilis* anti-alpha, global transcriptional regulator, Spx, in complex with the alpha C-terminal domain of RNA polymerase. Proc Natl Acad Sci USA 102:15839–15844

Newton GL, Fahey RC, Rawat M (2012) Detoxification of toxins by bacillithiol in *Staphylococcus aureus*. Microbiology 158:1117–1126

Newton GL, Leung SS, Wakabayashi JI, Rawat M, Fahey RC (2011) The DinB superfamily includes novel mycothiol, bacillithiol, and glutathione S-transferases. Biochemistry 50:10751–10760

Newton GL, Rawat M, La Clair JJ, Jothivasan VK, Budiarto T, Hamilton CJ, Claiborne A, Helmann JD, Fahey RC (2009) Bacillithiol is an antioxidant thiol produced in Bacilli. Nat Chem Biol 5:625–627

Nicolas P, Mader U, Dervyn E, Rochat T, Leduc A, Pigeonneau N, Bidnenko E, Marchadier E, Hoebeke M, Aymerich S, Becher D, Bisicchia P, Botella E, Delumeau O, Doherty G, Denham EL, Fogg MJ, Fromion V, Goelzer A, Hansen A, Hartig E, Harwood CR, Homuth G, Jarmer H, Jules M, Klipp E, Le Chat L, Lecointe F, Lewis P, Liebermeister W, March A, Mars RA, Nannapaneni P, Noone D, Pohl S, Rinn B, Rugheimer F, Sappa PK, Samson F, Schaffer M, Schwikowski B, Steil L, Stulke J, Wiegert T, Devine KM, Wilkinson AJ, van Dijl JM, Hecker M, Volker U, Bessieres P, Noirot P (2012) Condition-dependent transcriptome reveals high-level regulatory architecture in *Bacillus subtilis*. Science 335:1103–1106

Nielsen J, Hansen FG, Hoppe J, Friedl P, von Meyenburg K (1981) The nucleotide sequence of the atp genes coding for the F0 subunits a, b, c and the F1 subunit delta of the membrane bound ATP synthase of *Escherichia coli*. Mol Gen Genet 184:33–39

Nielsen PK, Andersen AZ, Mols M, van der Veen S, Abee T, Kallipolitis BH (2012) Genome-wide transcriptional profiling of the cell envelope stress response and the role of LisRK and CesRK in *Listeria monocytogenes*. Microbiology 158:963–974

Oden KL, Gladysheva TB, Rosen BP (1994) Arsenate reduction mediated by the plasmid-encoded ArsC protein is coupled to glutathione. Mol Microbiol 12:301–306

Ohki R, Giyanto Tateno K, Masuyama W, Moriya S, Kobayashi K, Ogasawara N (2003) The BceRS two-component regulatory system induces expression of the bacitracin transporter, BceAB, in *Bacillus subtilis*. Mol Microbiol 49:1135–1144

Ollinger J, Song KB, Antelmann H, Hecker M, Helmann JD (2006) Role of the Fur regulon in iron transport in *Bacillus subtilis*. J Bacteriol 188:3664–3673

Palm GJ, Khanh Chi B, Waack P, Gronau K, Becher D, Albrecht D, Hinrichs W, Read RJ, Antelmann H (2012) Structural insights into the redox-switch mechanism of the MarR/DUF24-type regulator HypR. Nucleic Acids Res

Pamp SJ, Frees D, Engelmann S, Hecker M, Ingmer H (2006) Spx is a global effector impacting stress tolerance and biofilm formation in *Staphylococcus aureus*. J Bacteriol 188:4861–4870

Parsonage D, Newton GL, Holder RC, Wallace BD, Paige C, Hamilton CJ, Dos Santos PC, Redinbo MR, Reid SD, Claiborne A (2010) Characterization of the N-acetyl-alpha-D-glucosaminyl l-malate synthase and deacetylase functions for bacillithiol biosynthesis in *Bacillus anthracis*. Biochemistry 49:8398–8414

Passalacqua KD, Bergman NH, Lee JY, Sherman DH, Hanna PC (2007) The global transcriptional responses of Bacillus anthracis Sterne (34F2) and a Delta sodA1 mutant to paraquat reveal metal ion homeostasis imbalances during endogenous superoxide stress. J Bacteriol 189:3996–4013

Passalacqua KD, Varadarajan A, Weist C, Ondov BD, Byrd B, Read TD, Bergman NH (2012) Strand-specific RNA-seq reveals ordered patterns of sense and antisense transcription in *Bacillus anthracis*. PLoS ONE 7:e43350

Perego M, Glaser P, Minutello A, Strauch MA, Leopold K, Fischer W (1995) Incorporation of D-alanine into lipoteichoic acid and wall teichoic acid in *Bacillus subtilis*. Identification of genes and regulation. J Biol Chem 270:15598–15606

Peris-Bondia F, Latorre A, Artacho A, Moya A, D'Auria G (2011) The active human gut microbiota differs from the total microbiota. PLoS ONE 6:e22448

Persuh M, Mandic-Mulec I, Dubnau D (2002) A MecA paralog, YpbH, binds ClpC, affecting both competence and sporulation. J Bacteriol 184:2310–2313

Petersohn A, Brigulla M, Haas S, Hoheisel JD, Völker U, Hecker M (2001) Global analysis of the general stress response of *Bacillus subtilis*. J Bacteriol 183:5617–5631

Pietiainen M, Gardemeister M, Mecklin M, Leskela S, Sarvas M, Kontinen VP (2005) Cationic antimicrobial peptides elicit a complex stress response in *Bacillus subtilis* that involves ECF-type sigma factors and two-component signal transduction systems. Microbiology 151:1577–1592

Piggot PJ, Hilbert DW (2004) Sporulation of *Bacillus subtilis*. Curr Opin Microbiol 7:579–586

Pohl S, Tu WY, Aldridge PD, Gillespie C, Hahne H, Mader U, Read TD, Harwood CR (2011) Combined proteomic and transcriptomic analysis of the response of *Bacillus anthracis* to oxidative stress. Proteomics 11:3036–3055

Predich M, Nair G, Smith I (1992) *Bacillus subtilis* early sporulaiton genes *kinA*, *spo0F*, and *spo0A* are transcribed by the RNA polymerase containing sigma H. J Bacteriol 174:2771–2778

Price C (2011) General stress response in *Bacillus subtilis* and related Gram-positive bacteria. In: Storz G, Hengge R (eds) Bacterial stress response. ASM, Washington, DC

Price CW (2002) General stress response. In: Sonenchein AL, Hoch JA, Losick R (eds) *Bacillus subtilis* and its closest relatives: from genes to cells. ASM, Washinton, DC, pp 369–384

Prudhomme M, Attaiech L, Sanchez G, Martin B, Claverys JP (2006) Antibiotic stress induces genetic transformability in the human pathogen Streptococcus *pneumoniae*. Science 313:89–92

Rabinovich L, Sigal N, Borovok I, Nir-Paz R, Herskovits AA (2012) Prophage excision activates *Listeria* competence genes that promote phagosomal escape and virulence. Cell 150:792–802

Reder A, Albrecht D, Gerth U, Hecker M (2012) Cross-talk between the general stress response and sporulation initiation in *Bacillus subtilis*—the sigma(B) promoter of *spo0E* represents an AND-gate. Environ Microbiol 14:2741–2756

Reder A, Gerth U, Hecker M (2012) Integration of sigmaB activity into the decision-making process of sporulation initiation in *Bacillus subtilis*. J Bacteriol 194:1065–1074

Reder A, Hoper D, Gerth U, Hecker M (2012) Contributions of individual sigmaB-dependent general stress genes to oxidative stress resistance of *Bacillus subtilis*. J Bacteriol 194:3601–3610

Reder A, Hoper D, Weinberg C, Gerth U, Fraunholz M, Hecker M (2008) The Spx paralogue MgsR (YqgZ) controls a subregulon within the general stress response of *Bacillus subtilis*. Mol Microbiol 69:1104–1120

Reder A, Pother DC, Gerth U, Hecker M (2012d) The modulator of the general stress response, MgsR, of *Bacillus subtilis* is subject to multiple and complex control mechanisms. Environ Microbiol 14:2838–2850

Renzoni A, Andrey DO, Jousselin A, Barras C, Monod A, Vaudaux P, Lew D, Kelley WL (2011) Whole genome sequencing and complete genetic analysis reveals novel pathways to glycopeptide resistance in *Staphylococcus aureus*. PLoS ONE 6:e21577

Rietkotter E, Hoyer D, Mascher T (2008) Bacitracin sensing in *Bacillus subtilis*. Mol Microbiol 68:768–785

Ritz D, Beckwith J (2001) Roles of thiol-redox pathways in bacteria. Annu Rev Microbiol 55:21–48

Rochat T, Nicolas P, Delumeau O, Rabatinova A, Korelusova J, Leduc A, Bessieres P, Dervyn E, Krasny L, Noirot P (2012) Genome-wide identification of genes directly regulated by the pleiotropic transcription factor Spx in *Bacillus subtilis*. Nucleic Acids Res

Rogstam A, Larsson JT, Kjelgaard P, von Wachenfeldt C (2007) Mechanisms of adaptation to nitrosative stress in *Bacillus subtilis*. J Bacteriol 189:3063–3071

Sauer RT, Baker TA (2011) AAA+ proteases: ATP-fueled machines of protein destruction. Annu Rev Biochem 80:587–612

Scharf C, Riethdorf S, Ernst H, Engelmann S, Volker U, Hecker M (1998) Thioredoxin is an essential protein induced by multiple stresses in *Bacillus subtilis*. J Bacteriol 180:1869–1877

Schimel J, Balser TC, Wallenstein M (2007) Microbial stress-response physiology and its implications for ecosystem function. Ecology 88:1386–1394

Schlothauer T, Mogk A, Dougan DA, Bukau B, Turgay K (2003) MecA, an adaptor protein necessary for ClpC chaperone activity. Pro Natl Acad Sci USA 100:2306–2311

Seredick SD, Spiegelman GB (2007) *Bacillus subtilis* RNA polymerase recruits the transcription factor SpoOA approximately P to stabilize a closed complex during transcription initiation. J Mol Biol 366:19–35

Shi J, Mukhopadhyay R, Rosen BP (2003) Identification of a triad of arginine residues in the active site of the ArsC arsenate reductase of plasmid R773. FEMS Microbiol Lett 227:295–301

Shi J, Vlamis-Gardikas A, Aslund F, Holmgren A, Rosen BP (1999) Reactivity of glutaredoxins 1, 2, and 3 from *Escherichia coli* shows that glutaredoxin 2 is the primary hydrogen donor to ArsC-catalyzed arsenate reduction. J Biol Chem 274:36039–36042

Shin JH, Brody MS, Price CW (2010) Physical and antibiotic stresses require activation of the RsbU phosphatase to induce the general stress response in *Listeria monocytogenes*. Microbiology 156:2660–2669

Silhavy TJ, Kahne D, Walker S (2010) The bacterial cell envelope. Cold Spring Harb Perspect Biol 2:a000414

Solomon JM, Magnuson R, Srivastava A, Grossman AD (1995) Convergent sensing pathways mediate response to two extracellular competence factors in *Bacillus subtilis*. Genes Dev 9:547–558

Staron A, Finkeisen DE, Mascher T (2011) Peptide antibiotic sensing and detoxification modules of *Bacillus subtilis*. Antimicrob Agents Chemother 55:515–525

Storz G, Spiro S (2011) Sensing and responding to reactive oxygen and nitrogen species. In: Storz G, Hengge R (eds) Bacterial stress responses. ASM, Washington, DC, pp 157–173

Stragier P, Losick R (1990) Cascades of sigma factors revisited. Mol Microbiol 4:1801–1806

Stragier P, Losick R (1996) Molecular genetics of sporulation in *Bacillus subtilis*. Ann Rev Genet 30:297–341

Strauch MA (1993) AbrB, a transition state regulator. In: Sonenshein AL, Hoch JA, Losick R (eds) *Bacillus subtilis* and other gram-positive bacteria: physiology, biochemistry, and molecular biology. ASM, Washington, DC, pp 757–764

Strauch MA, Webb V, Speigelman B, Hoch JA (1990) The Spo0A protein of *Bacillus subtilis* is a repressor of the *abrB* gene. Proc Natl Acad Sci USA 87:1801–1805

Suntharalingam P, Senadheera MD, Mair RW, Levesque CM, Cvitkovitch DG (2009) The LiaFSR system regulates the cell envelope stress response in *Streptococcus mutans*. J Bacteriol 191:2973–2984

Tan YP, Giffard PM, Barry DG, Huston WM, Turner MS (2008) Random mutagenesis identifies novel genes involved in the secretion of antimicrobial, cell wall-lytic enzymes by *Lactococcus lactis*. Appl Environ Microbiol 74:7490–7496

Tanous C, Soutourina O, Raynal B, Hullo MF, Mervelet P, Gilles AM, Noirot P, Danchin A, England P, Martin-Verstraete I (2008) The CymR regulator in complex with the enzyme CysK controls cysteine metabolism in *Bacillus subtilis*. J Biol Chem 283:35551–35560

Teplyakov A, Pullalarevu S, Obmolova G, Doseeva V, Galkin A, Herzberg O, Dauter M, Dauter Z, Gilliland GL (2004) Crystal structure of the YffB protein from *Pseudomonas aeruginosa* suggests a glutathione-dependent thiol reductase function. BMC Struct Biol 4:5

Thackray PD, Moir A (2003) SigM, an extracytoplasmic function sigma factor of *Bacillus subtilis*, is activated in response to cell wall antibiotics, ethanol, heat, acid, and superoxide stress. J Bacteriol 185:3491–3498

Tortosa P, Dubnau D (1999) Competence for transformation: a matter of taste. Curr Opin Microbiol 2:588–592

Tortosa P, Logsdon L, Kraigher B, Itoh Y, Mandic-Mulec I, Dubnau D (2001) Specificity and genetic polymorphism of the *Bacillus* competence quorum-sensing system. J Bacteriol 183:451–460

Tran LS, Nagai T, Itoh Y (2000) Divergent structure of the ComQXPA quorum-sensing components: molecular basis of strain-specific communication mechanism in *Bacillus subtilis*. Mol Microbiol 37:1159–1171

Turgay K, Hamoen LW, Venema G, Dubnau D (1997) Biochemical characterization of a molecular switch involving the heat shock protein ClpC, which controls the activity of ComK, the competence transcription factor of *Bacillus subtilis*. Genes Dev 11:119–128

Turlan C, Prudhomme M, Fichant G, Martin B, Gutierrez C (2009) SpxA1, a novel transcriptional regulator involved in X-state (competence) development in *Streptococcus pneumoniae*. Mol Microbiol 73:492–506

van Vliet AH, Baillon ML, Penn CW, Ketley JM (1999) Campylobacter jejuni contains two fur homologs: characterization of iron-responsive regulation of peroxide stress defense genes by the PerR repressor. J Bacteriol 181:6371–6376

Veiga P, Bulbarela-Sampieri C, Furlan S, Maisons A, Chapot-Chartier MP, Erkelenz M, Mervelet P, Noirot P, Frees D, Kuipers OP, Kok J, Gruss A, Buist G, Kulakauskas S (2007) SpxB regulates O-acetylation-dependent resistance of *Lactococcus lactis* peptidoglycan to hydrolysis. J Biol Chem 282:19342–19354

Verghese B, Lok M, Wen J, Alessandria V, Chen Y, Kathariou S, Knabel S (2011) *comK* prophage junction fragments as markers for *Listeria monocytogenes* genotypes unique to individual meat and poultry processing plants and a model for rapid niche-specific adaptation,
  biofilm formation, and persistence. Appl Environ Microbiol 77:3279–3292
Vijay K, Brody MS, Fredlund E, Price CW (2000) A PP2C phosphatase containing a PAS
  domain is required to convey signals of energy stress to the sigmaB transcription factor of
  Bacillus subtilis. Mol Microbiol 35:180–188
Wang C, Fan J, Niu C, Wang C, Villaruz AE, Otto M, Gao Q (2010) Role of spx in biofilm formation of *Staphylococcus epidermidis*. FEMS Immunol Med Microbiol 59:152–160
Weinrauch Y, Penchev R, Dubnau E, Smith I, Dubnau D (1990) A *Bacillus subtilis* regulatory
  gene product for genetic competence and sporulation resembles sensor protein members of
  the bacterial two-component signal-transduction systems. Genes Dev 4:860–872
Wolf D, Kalamorz F, Wecke T, Juszczak A, Mader U, Homuth G, Jordan S, Kirstein J, Hoppert
  M, Voigt B, Hecker M, Mascher T (2010) In-depth profiling of the LiaR response of *Bacillus
  subtilis*. J Bacteriol 192:4680–4693
Zhang Y, Nakano S, Choi SY, Zuber P (2006) Mutational analysis of the *Bacillus subtilis* RNA
  polymerase alpha C-terminal domain supports the interference model of Spx-dependent
  repression. J Bacteriol 188:4300–4311
Zhang Y, Zuber P (2007) Requirement of the zinc-binding domain of ClpX for Spx proteolysis in
  Bacillus subtilis and effects of disulfide stress on ClpXP activity. J Bacteriol 189:7669–7680
Zuber P (2004) Spx-RNA polymerase interaction and global transcriptional control during oxida-
  tive stress. J Bacteriol 186:1911–1918
Zuber P, Chauhan S, Pilaka P, Nakano MM, Gurumoorthy S, Lin AA, Barendt SM, Chi BK,
  Antelmann H, Mader U (2011) Phenotype enhancement screen of a regulatory *spx* mutant
  unveils a role for the *ytpQ* gene in the control of iron homeostasis. PLoS ONE 6:e25066

Index

P. Zuber, *Function and Control of the Spx-Family of Proteins Within the Bacterial Stress Response*, SpringerBriefs in Microbiology, DOI: 10.1007/978-1-4614-6925-4,
© The Author(s) 2013